Nucleic acids sequencing

a practical approach

TITLES PUBLISHED IN
THE
PRACTICAL APPROACH
SERIES

Series editors:
Dr D Rickwood
Department of Biology, University of Essex
Wivenhoe Park, Colchester, Essex CO4 3SQ, UK
Dr B D Hames
Department of Biochemistry, University of Leeds
Leeds LS2 9JT, UK

Affinity chromatography
Animal cell culture
Antibodies I & II
Biochemical toxicology
Biological membranes
Carbohydrate analysis
Cell growth and division
Centrifugation (2nd Edition)
Computers in microbiology
DNA cloning I, II & III
Drosophila
Electron microscopy
in molecular biology
Fermentation
Gel electrophoresis of nucleic acids
Gel electrophoresis of proteins
Genome analysis
HPLC of small molecules
HPLC of macromolecules
Human cytogenetics
Human genetic diseases
Immobilised cells and enzymes
Iodinated density gradient media
Light microscopy in biology
Liposomes
Lymphocytes
Lymphokines and interferons
Mammalian development
Medical bacteriology
Medical mycology
Microcomputers in biology

Microcomputers in physiology
Mitochondria
Mutagenicity testing
Neurochemistry
Nucleic acid and
protein sequence analysis
Nucleic acid hybridisation
Nucleic acids sequencing
Oligonucleotide synthesis
Photosynthesis:
energy transduction
Plant cell culture
Plant molecular biology
Plasmids
Prostaglandins
and related substances
Protein function
Protein sequencing
Protein structure
Proteolytic enzymes
Ribosomes and protein synthesis
Solid phase peptide synthesis
Spectrophotometry
and spectrofluorimetry
Steroid hormones
Teratocarcinomas
and embryonic stem cells
Transcription and translation
Virology
Yeast

Nucleic acids sequencing

a practical approach

Edited by
C J Howe

Department of Biochemistry,
University of Cambridge,
Tennis Court Road,
Cambridge CB2 1QW, UK

E S Ward

MRC Laboratory of Molecular Biology,
Hills Road,
Cambridge CB2 2QH, UK

—at—
OXFORD UNIVERSITY PRESS
Oxford New York Tokyo

IRL Press
Eynsham
Oxford
England

British Library Cataloguing in Publication Data

Nucleic acids sequencing
  1. Nucleic acids. Sequences
I. Howe, C.J. II. Ward, E.S.
III. Series
547.7'9

Library of Congress Cataloging in Publication Data

Nucleic acids sequencing: a practical approach/edited by C.J.Howe, E.S.Ward.
(Practical approach series)
  Includes bibliographies and index.
  1. Nucleotide sequence—Technique. I. Howe, C.J. II. Ward, E.S. III. Series.
[DNLM: 1. Base Sequence. 2. Computers. 3. DNA—analysis.
4. RNA—analysis. QU 58 N9653]
QP625.N89N835   1989   574.87'328—dc20   89-11227
ISBN 0 19 963056 9
ISBN 0 19 963057 7 (pbk.)

Previously announced as:
ISBN 1 85221 102 4 (hardbound)
ISBN 1 85221 103 2 (softbound)

Front cover illustration drawn from an original kindly supplied by
Ms C.Brown and Dr A.Bankier.

Typeset and printed by Information Press Ltd, Oxford, England

Preface

The methods of Sanger, Maxam and Gilbert for rapid DNA sequence determination have formed the cornerstone of the sequencing techniques used today. Improvements in the technology over the last decade (including the availability of better enzymes, higher quality reagents and more suitable equipment) have led to a large increase in the size of sequencing projects that people are prepared to undertake, and the accumulation of huge amounts of data. This has given a strong impetus to the development of automated sequencing and of course necessitated the refinement of computer hardware and software to handle the sequences generated. The methodology has spilled over, into RNA sequencing for example, and the applications into areas way outside the confines of molecular biology, such as archaeology and forensic work. This book aims to provide practical advice both for those who have not yet tackled sequencing and those who have, but might want to update their technology. We believe that there are very few books available that cover the whole range of nucleic acids sequencing, and we hope that this one will satisfy that need.

The first two chapters are concerned with what is probably the commonest type of sequencing project undertaken—the use of single-stranded DNA phage and related vectors in determining the sequence of a cloned piece of DNA using the chain-termination method. In the early stages, these projects are often fraught with problems which, although trivial, seem insurmountable at the time, and we hope that Chapter 3 will help in their resolution. Chapter 4 expands the techniques to other templates, such as double-stranded plasmid DNA and DNA produced by the polymerase chain reaction (which, due to its rapid and diverse applications, is becoming an increasingly important technique in molecular biology). For some projects, and some laboratories, the chemical sequencing method may be more suitable, and this, together with the use of metre-long gels (which may be useful to other sequencers too) is described in Chapter 5. A comprehensive coverage of RNA fingerprint analysis and sequencing is given in Chapter 6. The applications of computer systems, and the software available, are outlined in Chapter 7. This is the area that is perhaps likely to be least familiar to the users of this book, but is probably as important to the success of the project as the sequencing itself. Chapter 8 continues the discussion of automated sequencing (which began in Chapter 2) by a description of the use of fluorescent labels in the automation of the gel running and reading process.

We have asked the authors to warn, wherever possible, of the pitfalls in the techniques they describe. We have aimed to keep each chapter complete and self-contained, to minimize the need to refer to other chapters in the middle of a protocol, even where similar (but almost invariably non-identical) recipes are given by different authors for related manipulations. Aware that small and apparently insignificant changes in a protocol can readily affect its success, we have not attempted to 'unify' the recommended protocols. Finally, we would like to record our grateful thanks to the staff at IRL Press for their help and encouragement in compiling this book, and to the contributors for their willingness to participate.

C.J.Howe
E.S.Ward

Contributors

A.T.Bankier
MRC Laboratory of Molecular Biology, Hills Road, Cambridge CB2 2QH, UK

R.F.Barker
Innovation Centre, Astromed Ltd, Unit 6, Cambridge Science Park, Milton Road, Cambridge CB4 4GS, UK

B.G.Barrell
MRC Laboratory of Molecular Biology, Hills Road, Cambridge CB2 2QH, UK

M.J.Bishop
Computer Laboratory, University of Cambridge, Pembroke Street, Cambridge CB2 3QG, UK

C.Heiner
Applied Biosystems, 850 Lincoln Centre Drive, Foster City, CA 94404, USA

C.J.Howe
Department of Biochemistry, University of Cambridge, Tennis Court Road, Cambridge CB2 1QW, UK

T.Hunkapiller
Division of Biology, California Institute of Technology, Pasadena, CA 91125, USA

G.Krupp
Institut für Allgemeine Mikrobiologie, Christian-Albrechts-Universität, Olshausenstr. 40, D-2300, Kiel, FRG

J.Messing
Waksman Institute, Rutgers University, Piscataway, NJ 08855, USA

G.Murphy
Institute of Plant Science Research, Cambridge Laboratory, Maris Lane, Trumpington, Cambridge CB2 2JB, UK

E.Stackebrandt
Institut für Allgemeine Mikrobiologie, Christian-Albrechts-Universität, Olshausenstr. 40, D-2300, Kiel, FRG

D.A.Stahl
Department of Veterinary Pathobiology, College of Veterinary Medicine, University of Illinois, 2001 South Lincoln Avenue, Urbana, IL 61801, USA

E.S.Ward
MRC Laboratory of Molecular Biology, Hills Road, Cambridge CB2 2QH, UK

Contents

Abbreviations

BB	Bromophenol blue
Bis	N,N'-methylene-bis-acrylamide
BSA	Bovine serum albumen
CA	Cellulose acetate
Ctab	Cetyl triethylammonium bromide
dc7GTP	2'-deoxy-7-deazaguanosine 5'-triphosphate
(d)dNTP	(Di)deoxynucleoside triphosphate
DEAE	Diethylaminoethyl
DIGE	Direct agarose gel electrophoresis
DMS	Dimethyl sulphate
DMSO	Dimethyl sulphoxide
DTT	Dithiothreitol
EDTA	Ethylenediaminetetraacetic acid
HVE	High voltage electrophoresis
IPTG	Isopropylthiogalactoside
NTA	Nitrilotriacetic acid
ORF	Open reading frame
PCR	Polymerase chain reaction
PEG	Polyethylene glycol
RF	Replicative form
rtn	Reaction
SDS	Sodium dodecyl sulphate
SS	Single-stranded
TdT	Terminal deoxynucleotide transferase
TEMED	N,N,N',N'-tetramethylethylenediamine
TLC	Thin layer chromatography
XC	Xylene cyanol FF
Xgal	5-bromo-4-chloro-3-indolyl-β-D-galactoside

CHAPTER 1

The use of single-stranded DNA phage in DNA sequencing

JOACHIM MESSING and ALAN T.BANKIER

1. INTRODUCTION

1.1 The life cycle

The filamentous bacteriophage presents a unique mode of cloning. Double-stranded foreign DNA can be cloned into the replicative form (RF) phage DNA and, upon transformation, only one of the two strands is packaged into the viral coat (1). Therefore, strand separation and the cloning of DNA fragments are combined to yield large quantities of pure single-stranded DNA. Such a phage is M13, which used the F-pili of *Escherichia coli* as entry sites. Infection therefore requires the induction of the *tra* function of the F sex plasmid which is repressed in the absence of aeration. One *tra* mutation, *tra*D36, is a leaky mutation that reduces conjugation by a factor of 10^{-5}, but still gives wild-type levels of phage titre (2). This mutation is necessary for reasons of biological containment.

In the early stages of recombinant DNA work, guidelines were established to define biological containment. A particular concern at the time was the possible escape of recombinant DNA into the environment. Since many forms of *E.coli* grow in the human gut, a health risk was also feared. The anaerobic conditions in the gut can be expected to keep pili formation within guideline limits, but the containment of conjugative plasmids in the environment creates a different problem because they can find an ecological niche by transfer to other bacteria. For example, the spread of antibiotic-resistance genes is facilitated by conjugative plasmids, a phenomenon that presents a major problem to hospitals and farms. Accordingly, the NIH guidelines require the use of non-conjugative plasmids as vectors. For this reason, the F sex plasmid is not used as a vector and bacterial hosts containing one are excluded. The *tra*D36 mutation, discussed above, converts the F sex plasmid into a conjugation-deficient plasmid. It is interesting that M13 infection reduces conjugation of the F sex plasmid even more efficiently. Cell infection reduces conjugation by a factor of 10^6, and in combination with the *tra*D35 mutation, by 10^{11} (3).

After the phage has injected its circular single-stranded DNA, called the plus strand, into the bacteria, host cell functions can convert the single-stranded DNA into a double-stranded DNA form, also called the parental RF or replicative form. A specific site on the plus strand, the origin, forms a hairpin and acts as a weak promoter. The RNA produced serves as primer for the synthesis of the minus strand. The minus strand is then transcribed to produce all the viral products. One of them, the gene IIp(roduct), recognizes another site at the origin, where it introduces a nick into the plus strand

1

at nucleotide 5780 (4). The same protein can also act as a topoisomerase to seal the nick. The nick provides the 3'-OH end necessary for DNA polymerase to copy the minus strand. This synthesis replaces the old plus strand, giving a structure that appears under the electron microscope as a rolling circle. Another nick is necessary to separate the old and new plus strand. This is carried out by a new gene IIp molecule, while the old one closes the replaced strand to a ring. Therefore, gene IIp is not an enzyme, but acts stoichiometrically. Since the released single-stranded ring can be reconverted into the RF form by host functions, about 100 RF molecules accumulate per cell (5).

Infected cells can therefore be used to prepare RF like any multicopy plasmid, the resulting DNA can be treated with restriction enzymes, and form chimaeric double-stranded circles with any other DNA molecule. Like plasmid vectors, RF DNA can be used to transform *E.coli* by the $CaCl_2$ technique, except that whereas plasmids require a selectable marker for the detection of transformed cells, RF gives rise to plaques. Unlike phage λ cos sites, RF cannot be packaged *in vitro*, which is not surprising because M13 packaging requires the single-stranded (ss) plus strand. However, highly competent *E.coli* cells can yield 10^8 plaques per microgram of RF. They are available from a number of suppliers.

As the number of RF molecules increases, another viral product, gene Vp, increases as well. Gene Vp has two functions: it reduces the translation of gene IIp, and it binds to the newly synthesized plus strands. This balances the production of RF molecules at two different steps, reducing the initiation of plus strand synthesis and the conversion of plus strand into RF molecules (*Figure 1*). The gene Vp is replaced by gene VIIIp in a step important to the secretion of the plus strand by the bacteria. Filamentous phage have a remarkable ability to leave the host without lysis. Under optimal conditions *E.coli* can keep producing M13 phage constantly, but cell division time increases from 20 min to more than 2 h. This has the advantage that, when grown on agar plates, infected cells surrounded by uninfected cells give the appearance of a plaque, which is important in titrating a phage solution and cloning new chimaeric phage. It should be noted that release of phage from host does not require the pili produced by the F sex plasmid. For example, if competent F^- cells are transformed with RF, phage production is normal, except that no plaques can form because secreted phage cannot infect. Transformed host mixed with F^+ bacteria can, however, be infected by phage, and plaques form when the mixture is plated under appropriate conditions. Experimental procedures can be devised that make use of easily distinguishable primary and secondary hosts.

Phage production may vary considerably, but titres of 10^{12} ml^{-1} or a few thousand per cell are quite normal, yielding about 10 μg in a 1-ml culture. Phage are counted as plaque forming units (p.f.u.). A titre of 3×10^{11} p.f.u. is equivalent to 1 μg of single-stranded M13 DNA. Since there are about 100 RF molecules and 20 times as many phage per cell, the theoretical yield of a 100-ml culture grown to about 2×10^9 cells ml^{-1} is 0.13 mg RF and 1.3 mg of ss DNA. This amplification of single-stranded DNA and its separation from cellular nucleic acids by intact cells constitutes a major purification step that increases the resolving power of hybridization and DNA sequencing techniques. It is, however, crucial that host cells do not lyse, because RNA fragments could be purified with the phage and act as random primers when the plus strand is used as a template in DNA synthesis. Some *E.coli* strains lyse more readily than others.

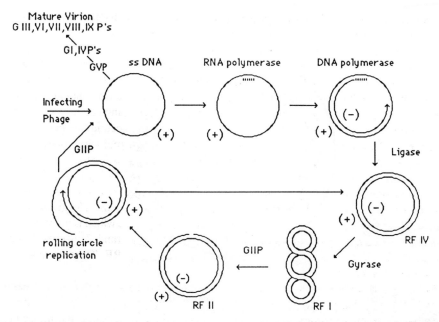

Figure 1. M13 life cycle. Explanations are given in the text. Figure courtesy of J.Vieira.

Care should also be taken to harvest infected cells by the end of the log phase because some lysis also occurs during the stationary phase. Therefore, fresh, rapidly dividing cultures are the preferred starting source.

1.2 Cloning into the RF

Rather than using a rare cutting restriction endonuclease like *Eco*RI, the first cloning into RF used *Bsu*I which cuts RF ten times. This had the advantage of directing the insertion to a non-essential site where no natural cloning sites exist. Today such a site could simply be made by site-directed mutagenesis; the result is the same. Cloning into the RF must occur within a very narrow region of the viral genome to ensure high titres of recombinant phage. This region is one of the two that contain no reading frame, but do contain other vital information (*Figure 2*). The smaller one (between genes VIII and III) contains an important structure for regulating the expression of coat proteins. Formally, it is a termination site for transcription and a promoter site for initiation of transcription. The other region has the same two properties. As a consequence, polycistronic messengers are synthesized that terminate in these regions opposite each other in the genome. There are additional transcriptional start sites along the minus strand, increasing the number of messenger molecules for distal gene products, all terminating in the two termination sites. Termination and start of transcription in the small region overlap and no transcription-free region is generated. The larger intergenic region between genes IV and II, however, has a transcription-free gap that contains the two origins, one each for plus and minus strand synthesis. Therefore, insertion must occur in the transcription-free gap without destruction of the origins. Although there are two *Bsu*I sites in this region, the first cloning led to insertion of DNA into the origin

3

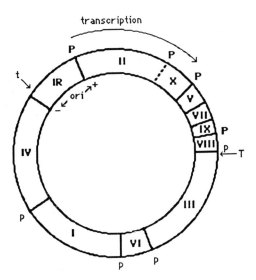

Figure 2. Circular genetic map of the filamentous phage genome. The genes are designated by Roman numerals. IR refers to the intergenic region, which contains the origin of replication for both the plus (+) and minus (−) strands. The direction of transcription is indicated. The more active promoters are designated by (**P**) and the less active by (**p**). The *rho* independent signal for the termination of transcription is indicated by (**T**) and the *rho* dependent termination signal by (**t**). The first cloning into M13 RF led to the insertion of the *lac* DNA into the *Bsu*I site at position 5868 (see also *Figure 3*). The resulting recombinant phage was called M13mp1. Figure is courtesy of J.Vieira.

for plus strand synthesis. The recombinant phage, M13mp1, however, gives titres near the wild-type levels.

A closer analysis of this insertion mutant (6) revealed two domains, A and B, within the intergenic region that are recognized by the gene IIp (product) (*Figure 3*). The site of the nick, at position 5780, is in domain A. Deletion of domain B reduces the phage titre 100-fold, making it useless as a cloning vector. Interestingly, the domain B function can be rescued. Domain B is neither an entry site of the gene IIp nor essential for strand selection, since these functions are retained in the absence of this domain, but it somehow enhances the efficiency of the gene IIp. The rescue comes in different forms, one by increasing the threshold of gene IIp and another by a single-site mutation in gene II. The threshold of gene IIp can be altered by a mutation in *cis* or *trans* that affects the translation of the gene II message. The *trans*-acting factor is encoded by the distal gene V. A single amino acid change in the amino-terminal region of the single-stranded DNA-binding protein encoded by gene V results in a failure to down-regulate translation of the gene II mRNA. This increases the levels of gene IIp about ten times, which is sufficient to overcome the effect of the insertion mutant of domain B. The mutation in *cis* effects a single base change in the 5′ leader sequence of the gene II mRNA. By contrast, gene IIp of M13mp1 is produced at wild-type levels, but contains a single amino acid change at residue 40. The altered M13mp1 gene IIp function rescues a mutant domain B. The lengthy discussion of these mutations becomes clear in the discussion of cloning single-stranded plasmids. At this point, it is sufficient to realize that domain B can be used as a cloning site within the RF of M13mp1 and all its derivatives.

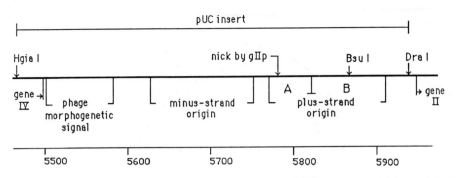

Figure 3. Map of the intergenic region of M13. The *Hgia*I−*Dra*I fragment provides all the *cis*-required functions for replication and has been cloned into pUC18 and 19 to produce the plage vectors pUC118 and 119 as discussed later. The nick introduced by gIIp at position 5780 (in domain A) and the insertion of the *lac*-DNA into the *Bsu*I site at position 5868 (in domain B) resulting in the M13mp phage vectors are marked. Further explanation is given in the text. Figure courtesy of J.Vieira.

Multiple cloning sites, as they were known in plasmids, could not be distributed all over the viral genome, but must be within domain B. Furthermore, since recombinant phage are detected by plaques rather than colonies, a drug-resistance marker used for plasmid-cloning vectors to screen recombinant plasmids cannot be applied to M13 cloning. To screen infected cells within a lawn of uninfected cells, a histochemical reaction is superior, such as the galactosidic cleavage of Xgal (5-bromo-4-chloro-3-indolyl-β-D-galactoside). Although the compound itself is colourless, its cleavage products are galactose and indigo, the latter giving rise to a dark blue colour. Since Xgal is taken up into cells, this reaction can occur when the enzyme β-galactosidase, a product of a structural gene of the *lac* operon, is synthesized. This enzyme offers two other advantages: its expression is regulated by a repressor that can be controlled, and a mutant enzyme that fails to tetramerize can be intracistronically complemented by a minigene.

The inducer IPTG (isopropylthiogalactoside) simulates the action of lactose by binding to the repressor. The inducer−repressor complex no longer binds to the *lac* operator, which leads to the expression of the *lacZ* gene encoding β-galactosidase. On the other hand, Xgal fails to bind to *lac* repressor and in the absence of IPTG no induction of the *lacZ* gene would occur. However, IPTG cannot be cleaved by β-galactosidase, leaving the concentration of the inducer unaltered. Therefore, IPTG and Xgal can be used to regulate a highly sensitive histochemical reaction.

For complementation the minigene requires only a small fraction of the *lac* operon provided that the entire operon is present either on the bacterial or F' chromosome. It requires the *lac* regulatory region and the amino-terminal region of the *lacZ* gene (7). Such a minigene, containing only 10% of the *lacZ* gene, was originally cloned into domain B of M13 RF and the recombinant named M13mp1. The F' factor of the host for propagating the phage carries the entire *lac* operon with two mutations; one is a small deletion of codons 11−41 of the *lacZ* gene. Since this is an in-frame deletion, a nearly full-length protein is produced, called the M15 protein because the mutation is likewise M15, which has nothing to do with the name M13 (8). The M15 protein is stable, but does not tetramerize, a requirement for enzyme function. When the minigene is expressed, the protein fragment produced can restore the M15 tetramerization. Although this intracistronic complementation results in less than 1/1000-fold β-galactosidase activity compared to wild-type, the histochemical reaction is so sensitive that the function of the minigene is faithfully reported. The other *lac* mutation on the F' episome is a single base change in the *lacI* promoter controlling the constitutive expression of *lac* repressor (9). The mutation, also known as *lacI*^{q}, leads to a 10-fold higher level of repressor, yielding about 100 repressor molecules per cell. Since the minigene also carries the *lac* operator but not the repressor gene, the multicopy RF would titrate the repressor very quickly. This can easily be tested. If a host with the wild-type *lac* operon is infected, cells give a full blue colour in the absence of an inducer. If a host with the *lacI*^{q} mutation is used, cells remain colourless in the absence of an inducer. Interestingly, long-incubated plates begin to show a blue ring around the plaques, indicating a repressor titration that agrees well with the estimates of 100 repressors and 100 RFs per cell. The addition of IPTG, however, produces a full colour reaction.

Besides the *lac* operon and the two mutations described above the F' factor carries two additional important markers. The *tra*D36 mutation has already been discussed; it is used for biological containment. The second is the operon for proline synthesis, *pro*AB. Defects in this region of the bacterial chromosome give rise to proline auxotrophs. Therefore, the absence of proline in the medium can be used to select for the presence of F' factor. The bacterial chromosome must of course have a deletion large enough to prevent homologous recombination between the F' *lac-pro* region and the chromosome. Because of all the described mutant markers, a standard host for M13mp vectors now carries the following phenotype: JM101 Δ(*lac pro*) *thi*-1 *sup*E F' [*pro* AB$^+$, *lacI*^{q}Z(ΔM15), *tra*D36] (3). Other mutations were added later, such as r$^-$ m$^+$ for cloning unmodified DNA and *rec*A for preventing recombination between repeated sequences. Various strains have also been tested for high efficiency transformation. Here, the primary and secondary host could come into play. The primary, which can be an F$^-$ host, could be used to obtain high efficiency transformation, and the secondary an F$^+$ strain for detection and propagation. The *sup*E mutation can be used selectively to propagate versions of the M13mp vectors that carry amber mutations.

1.3 The polylinker

Cloning of the *lac* minigene into the RF was easily detected by the formation of blue plaques under appropriate plating conditions. If any cleaved circular molecule joins with a restriction fragment produced by cleavage with a single enzyme, two recombinant DNA molecules can result that differ in the orientation of the insert. If the top strand of the fragment is called A and the bottom one B, in one case the A strand is linked to the plus strand of the phage, and in the other the B strand is linked to the plus strand of the vector. Since only the plus strand is packaged, one recombinant phage contains the A strand and the other the B strand of the cloned DNA fragment. In the case of the *lac* minigene, the coding strand becomes the minus strand, placing the *lac* promoter in tandem with the viral gene II promoter. Therefore, both strands of a double-stranded DNA are cloned separately. The cloning in two orientations can also be used in the C-test (see section 3.5.2) (10). Two M13 plus strands do not hybridize with each other, but if they contain two complementary DNA strands, hybridization can occur via the cloned strands, givine rise to a figure-of-eight structure.

After the *lac* minigene had been cloned into the RF, all subsequent cloning was done with the M13 *lac* phage for the convenience of histochemical screening of recombinants. At present the absence of Xgal cleavage or a change from blue to colourless is used as a screening method; this procedure is called insertional inactivation. Therefore, the M13 vector needs a cloning site at the beginning of the *lacZ* gene leading to the inactivation of the *lac* minigene, which results in colourless plaques. By means of site-directed chemical mutagenesis, an *Eco*RI site was constructed at position codon 5 of the *lacZ* gene (11). Although this resulted in a change from aspartic acid to asparagine, the new phage, M13mp2 still gave rise to blue plaques. However, *Eco*RI fragments cloned into this site inactivate the minigene, which now can be used to screen for the insertion of *Eco*RI fragments into the M13 RF. Although this should now provide all the components for cloning single-stranded DNA, it would be tedious to generate another M13 vector for every cloning site. Fortunately, when a *Hind*III linker was inserted into the *Eco*RI site in-frame, a phage called M13mp5 resulted, in which no inactivation of the minigene occurred (3). In this way, a series of vectors, M13mpn, where n is an integer, has been constructed that differ in the number and nature of the linker sites at codon 5 of the *lacZ* gene. Because of the use of multiple linkers, the resulting site is also called a polylinker or polycloning site (*Figure 4*).

An initial polylinker *Eco*RI-*Bam*HI-*Sal*I-*Pst*I-*Sal*I-*Bam*HI-*Eco*RI gave rise to M13mp7. Because of the symmetry around the *Pst*I site, a kanamycin-resistance gene was cloned into this site. The pUC counterpart of this phage was called pUC4K. It made it possible to cut out the drug-resistance gene with a battery of different enzymes, with the advantage that all the fragments have different restricted ends suitable for cloning elsewhere. This unit was called originally RSM, a *r*estriction cleavage *s*ites *m*obilizing element (12) and was used to make random in-frame insertions into the β-lactamase coding region (13). In addition, two other types of polylinkers were constructed. The one in M13mp7 did not permit the use of two cloning sites at the same time. That facility would have two advantages:

(i)   it would prevent religation of the vector DNA, and

(ii)   it fixes the orientation of the inserted fragment in a process called forced cloning (14).

Since the cloning of both single strands of a double-stranded DNA requires cloning in both orientations, two polylinkers were constructed with single sites in each orientation. The two resulting phages, M13mp8 and M13mp9, have been the basis for new derivatives with more and more sites. The present M13mp18 and M13mp19 are frequently used (15).

1.4 The universal primer

An important reason for using cloning vectors capable of producing single-stranded DNA was the ability to use this as a template for DNA sequencing using the dideoxynucleotide chain-termination method. The original sequencing technique was tested by determining the sequence of another single-stranded DNA phage, ϕX174, using double-stranded restriction fragments of RF as primers. It would be tedious to prepare a new primer for every sequencing reaction; the M13 vectors offer a simple

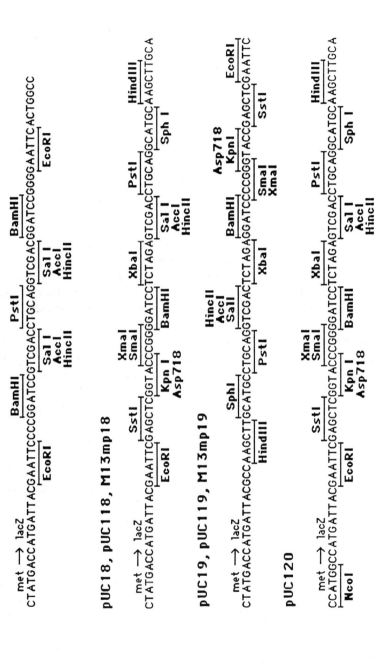

Figure 4. Common polylinkers in pUC and M13mp vectors showing the nucleotide sequences encoding the amino-terminal end of the modified *lacZ* gene. The ATG start site is marked. The ACG codon containing the start of the *Eco*RI site in the even-numbered vectors is the fifth codon. The AGC codon of the *Hin*dIII site in the odd-numbered vectors is the seventh codon. M13mp7 and pUC4 start with the *Eco*RI site which occurs twice in the polylinker. Insertion of the *Aph*II gene into the *Pst*I site of pUC4 led to pUC4K and gave rise to kanamycin resistance. This plasmid was the predecessor of both the even-numbered and odd-numbered polylinkers (12). Figure courtesy of J.Vieira.

solution. Since all fragments would be flanked by the same M13 sequence, only one primer is necessary for sequencing. It is complementary to the coding region of the *lacZ* gene, which is invariant. The primer initiates the sequencing reaction close to the polylinker, and all sequencing gels would have this start in common, but differ where the insert begins. Such a primer is therefore called a universal primer (16). With the advent of rapid chemical synthesis of DNA, a short oligonucleotide became available as a primer. Another universal primer that points away from the insert leads to the synthesis of the minus strand and has been used to make single-stranded specific hybridization probes (17).

1.5 **Shotgun sequencing**

Given the polylinker and the universal primer, another scheme became obvious. To utilize fully the speed of a universal primer, the DNA can be dissected into small overlapping fragments all cloned randomly into the phage. Sequencing is continued until overlapping information permits the reconstruction of a physical map of the original DNA (18). In particular the *Hin*dII site in M13mp7 proved to be versatile because any fragment with a blunt end could be cloned. Restriction fragments were initially used to sequence the complete genome of a plant virus, cauliflower mosaic virus, (19) and bacteriophage λ DNA (20). This process can be accelerated by cloning either DNaseI treated DNA (21,22) or DNA fragments produced by shearing (10,23−25). The process normally results in a high redundancy of data. Towards the end of a project, it may prove better to perform some selective cloning to close gaps separating established sequence information, but for large projects it is still the fastest approach.

1.6 **The progressive deletion method**

An alternative to shotgun sequencing is the preparation of ordered DNA templates by progressive removal of DNA ends. The DNA is treated with an exonuclease, reaction samples are withdrawn at intervals, and the degradation is stopped. The DNA samples are then cloned into a single-stranded DNA vector. DNA from recombinant phage from different sampling times is then subjected to gel electrophoresis, giving different molecular weights that confirm the extent of the deletion from a given end. Since all the ends can be sequenced with the universal primer, this technique should systematically connect all sequence information. This method has less redundancy than shotgun sequencing, but the cloning requires more labour. The exonuclease reaction must be tested and the recombinant phage electrophoresed; the overall procedure is more expensive and probably more suitable for small sequencing projects. Progressive deletion was first applied by Poncz *et al.* (26) using Bal31.

In the procedure employed in this chapter recombinant DNA is cleaved with two enzymes in the polylinker, on the primer annealing side of the insert. One produces a 3′-OH single-stranded end (*Pst*I) and the other a 3′-OH recessed end (*Xba*I). The linearized DNA is incubated with exonuclease III for various time periods (27). The 3′-OH overhang, but not the 3′-OH recessed end, is resistant to the progressive exonuclease III deletion. Samples taken at different times are treated with exonuclease VII to remove all single-stranded DNA, and the resultant double-stranded DNA ends are joined. After the ligated DNA is cloned, the recombinant phage is subjected to gel electrophoresis (15).

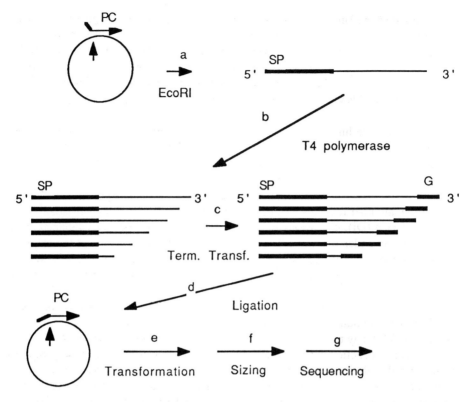

Figure 5. T4 deletion method. The oligonucleotide PC is annealed to the ss M13 template at the polylinker region between the inserted foreign DNA and the universal sequencing primer site (**SP**). The polyC tail of the oligonucleotide remains unpaired. The duplex DNA is cut with *Eco*RI. The primer is shortened and falls of (**a**). The linear single-stranded DNA is incubated with T4 polymerase to different extents (**b**). Reaction products are treated with terminal transferase to add polyG (**c**). Circular single-stranded DNAs are stabilized by hybridizing the oligonucleotide with the polyC to the ends of the circle. The nick is sealed by DNA ligase (**d**). The circles are transformed (**e**), and recombinant phage is sized (**f**) and sequenced (**g**).

Alternatively, by hybridizing an oligonucleotide to the single-stranded DNA at the cloning site, the ss DNA can be linearized at the 3′ end of the cloned DNA. Subsequently, the ss DNA is progressively degraded by the exonucleolytic activity of T4 polymerase (28). Before the ends can be joined, the 3′-OH end is tailed by terminal transferase, and the oligonucleotide that has the complementary end (left unpaired in the first reaction) bridges with the two ends and permits ligase to reseal them. The reaction products are again used to transform *E.coli*, and the molecular weights of the recovered recombinant DNAs are determined by gel electrophoresis (*Figure 5*).

1.7 **Plasmid−phage chimaeric vectors**

Although plasmid vectors like pBR322 are versatile and sufficient for most requirements, the histochemical screen of M13 recombinants with a *lac* minigene and the presence of the two reverse polylinkers would be equally useful within a plasmid system. For this reason, chemically induced single base changes as well as deletions were used to free pBR322 from cloning sites that could interfere with the use of the polylinker. This work started at the University of California at Davis, hence, the name pUC. During the search for the proper derivatives, it was noticed that the pUC derivatives titrated the *lac*I^q mutation. The plasmid evidently gives a higher copy number than RF. These new vectors were not only useful for screening recombinants by a simple histochemical reaction, but also yielded high levels of plasmid DNA without the usual chloramphenicol amplification. Yields were so high that CsCl gradients became unnecessary in plasmid cloning (12).

Because of these high yields and the stability of recombinant pUC plasmids, they and their commercial derivatives are now very widely used. DNA fragments were frequently passed back and forth between M13 and pUC vectors. However, a number of years ago it was observed that plasmids can also be packaged as single-stranded DNA (29). When plasmids contain the viral intergenic region, containing the origin of DNA replication, all other functions can be provided in *trans*. This is not surprising because defective phage (DI), comprising only the intergenic region and part of gene II, have been observed under the electron microscope (30). However, even under the best conditions, the yields of these defective phage or plasmid−phage hybrids (plage) were at least ten times lower. The low yields of plage, in particular, hampered preparation of template for DNA sequencing (31).

This low yield is somewhat unexpected. If two such origins are in the same cell, the titre of each could be expected to drop by half. In practice, a greater reduction (about 100-fold) may result from the infection of a host carrying a plasmid with a phage origin. Such a reduction is also called interference. If mutant variants of the phage can be recovered with increased titres, they are called interference-resistant mutants (4).

Normally, the plasmid would replicate from its own origin, but when gene IIp is made by a phage, it can act in *trans* on the plasmid, initiating a 'rolling circle' replication. The plage origin is then in competition with the phage origin. To overcome this competition and favour the replication of plage as a defective phage, two steps were taken. The first stems from the characterization of the M13mp1 phage (6), which contains a mutant origin where domain B is interrupted. Although a mutant gene IIp can compensate for this mutation as described above, interestingly it still prefers a wild-type origin. This can easily be tested by infecting hosts with plage carrying wild-type and mutant origins. The titre of M13mp1 is much lower than that of plage, which is about 10% of the normal wild-type phage levels. Although the plage replicates preferentially as a defective phage, the low titre of the helper phage rapidly leads to the limitation of viral gene products necessary to package the defective phage (Vieira and Messing, unpublished results). To amplify the helper phage, a plasmid origin is added to permit plasmid replication of phage RF. The resulting helper phage is called M13KO7 (32); it is derived from wild-type M13 phage by the insertion into domain B of the pACYC plasmid origin (which is compatible with the pUC origin) and of the APH I gene from pUC4K (see above) to select for helper phage growth in the presence of kanamycin. The second step taken was the exchange of the wild-type gene II for the one from M13mp1.

The intergenic region of wild-type M13 is contained within pUC118, pUC119 and pUC120. When host cells carrying these plasmids are infected with M13KO7, the pUC plasmids adopt a 'rolling circle' and the helper phage a 'plasmid' mechanism of replication. Such infections can result in titres of 5×10^{11} colony-forming units (c.f.u.)

per ml, 10 times higher than earlier helper phage. These plage yields permit DNA sequencing. Because single-stranded DNA replication now occurs only during the presence of the helper phage, larger inserts can also be packaged as single-stranded DNA that might have been less stable in M13 vectors.

2. MATERIALS

For a list of materials, see *Table 1*.

3. PROCEDURES

3.1 Growth and maintenance of strains

Infective phage particles, frozen in broth at $-20°C$ are stable for many years. Bacterial cells from an infected culture are removed by low speed centrifugation ($6000-8000$ g, 10 min). The supernatant, containing the free phage, can be frozen without osmotic agents; this also applies to the plage solution. Inoculation with a few ice crystals under sterile conditions is enough to infect a growing culture of the appropriate host.

Host cells can also be kept in a rich medium. However, to prevent bursting of the cells during freezing, the culture medium should contain 7% dimethyl sulphoxide (DMSO) or 15% glycerol. Under these conditions long-term storage is possible at $-70°C$. For the short term (a few years), they are stored at $-20°C$ in $40-50$% glycerol. For routine use, a slant culture can be kept in the refrigerator for several months. Strains should be routinely tested for markers. For example, test for proline auxotrophy in a minimal medium to ensure the presence of the F′ factor. For details of the growth, maintenance and purification of bacterial and viral strains, see Hackett *et al.* (33). It is always advisable to streak out cells repeatedly for single colonies, and dilute phage solution to single plaques.

This precaution is particularly important with a chimaeric helper phage such as M13KO7 described above (32). Since it carries the kanamycin-resistance gene from pUC4K and the replication origin from plasmid p15A, it can lose part of its DNA and replicate independently. This instability should not be a problem if the following procedure is adhered to.

(i) Streak an aliquot of M13KO7 from stock solution onto a fresh YT agar plate with a sterile wire loop, as is done with bacterial cultures.

(ii) Prepare a fresh culture of JM101 (or equivalent strain), which has been grown to an OD_{600} of at least 0.8 by adding 4 ml of YT soft agar to 0.5 ml of culture.

(iii) Pour the mixture onto the streaked plate from the side of the plate where the phage solution is diluted to the side where the streaking began. Wait for the soft agar to harden, and incubate at 37°C for $6-12$ h.

(iv) Inoculate a $2-3$-ml culture of YT containing kanamycin (70 μg ml^{-1}) with a single plaque from the overnight plate. Grow this culture overnight ($12-16$ h) with moderate agitation.

(v) Remove infected cells by centrifugation ($8000-10\ 000\ g$, 10 min). Transfer the phage supernatant to a sterile tube, determine the titre ($\sim 10^{12}$ p.f.u. ml^{-1}), and store as described above.

Table 1. Materials.

M9 salts (10×)
  Na_2HPO_4 60 g
  KH_2PO_4 30 g
  NaCl   5g
  NH_4Cl 10g
Dissolve in 1 litre of distilled water and autoclave; keep at room temperature.

M9 minimal medium
  M9 salts (10×)   100 ml
  $MgSO_4 \cdot 7H_2O$ (1 M) 1 ml
  Glucose (20%)   10 ml
  Vitamin B1 (1%)   1 ml
  $CaCl_2$ (10 mM)   10 ml
Dissolve solids in distilled water, filter sterilize, and add M9 salts to a final volume of 1 litre of water; keep at room temperature.

YT medium (rich medium)
  Bactotryptone   16 g
  Bacto yeast extract 10 g
  NaCl   5 g
  Glucose (20%)   10 ml
  Vitamin B1 (1%) 1 ml
Dissolve in 1 litre of distilled water, autoclave, and add filter-sterilized glucose and thiamine; keep at room temperature.

B broth (rich medium)
  Bactotryptone 10 g
  NaCl   8 g
Dissolve in 1 litre of distilled water, autoclave, add 1 ml of filter-sterilized 1% vitamin B1; keep at room temperature.

Lysis solution I
  25% sucrose
  25 mM Tris−HCl (pH 8.0)
  20 mM Na_2EDTA
Prepare immediately before use.

Lysis solution II
  0.2 M NaOH
  1% SDS
Prepare immediately before use.

TE
  20 mM Tris−HCl (pH 8.0)
  0.2 mM Na_2EDTA

NaOAc solution
  For pH 4.7 adjust with glacial acetic acid
  For pH 7.0 adjust with NaOH
  Mix slowly because of exothermic reaction.

Table 1. (continued)

Ctab QN (quaternary ammonium) butanol
(i)   equlibrate 150 ml of n-butanol with 150 ml of glass-distilled water by shaking in a separating funnel.
(ii)   Separate the phases and add 1 g of Ctab (hexadecyltrimethylammoniumbromide) to 100 ml of the butanol fraction.
(iii)   Shake well with 100 ml of the equilibrated aqueous phase [add 50 μl of antifoam A (Sigma) to prevent the formation of emulsions].
(iv)   Allow the solutions to separate overnight and bottle them separately (QN butanol and QN aqueous solution); they can be stored at room temperature for use.

ExoIII buffer
  50 mM Tris$-$HCl (pH 8)
  5 mM MgCl$_2$
  1 mM DTT

ExoVII buffer (10×)
  500 mM KH$_2$PO$_4$ (pH 7)
  80 mM Na$_2$EDTA
  10 mM DTT

Ligation buffer (10×)
  250 mM Tris$-$HCl (pH 7.5)
  100 mM MgCl$_2$
  25 mM hexamine cobalt chloride
  5 mM spermidine

Loading buffer
  0.05% (w/v) bromophenol blue
  0.2 M Na$_2$EDTA, pH 8.3
  50% glycerol

T4 DNA pol buffer (10×)
  0.33 M Tris/OAc (pH 7.9)
  0.66 M KOAc
  0.10 M Mg(OAc)$_2$

Tris$-$borate buffer:
  Tris base 363.3 g
  Boric acid 185.5 g
  Na$_2$EDTA 17.5 g
  Dissolve in 3 litres of H$_2$O

Ethidium staining bath
  Ethidium bromide (0.5 mg l^{-1})
  1 mM Na$_2$EDTA (pH 8.3)

SSC (1×)
  0.15 M NaCl
  0.015 M sodium citrate

TFB medium
  100 mM KCl
(ultrapure)   7.4 g
  45 mM MnCl$_2 \cdot$H$_2$O   8.9 g

10 mM $CaCl_2 \cdot 2H_2O$	1.5 g	
3 mM $(NH_2)_6CoCl_3$	0.8 g	
0.5 M K-MES (pH 6.3)	20 ml	

(i)   Prepare a 0.5 M K-MES solution, and adjust the pH to 6.3 with KOH.
(ii)   Prepare a solution of 10 mM K-MES from stock, and add the salts as solids.
(iii)   Adjust volume with 1 litre of distilled water.
(iv)   Sterilize by filtration through a rinsed (H_2O) 0.22-μm filter.
(v)   Store aliquots at 4°C.

DnD solution
   1 M DTT 1.53 g
   DMSO 9.0 ml of a 90% solution
   10 mM KOAc 100 μl of 1 M stock
Dissolve in 10 ml of distilled water.

RF1 solution

100 mM RbCl	12.0 g
50 mM $MnCl_2 \cdot 4H_2O$	9.9 g
30 mM KOAc 30 ml of 1 M	
10 mM $CaCl_2 \cdot H_2O$	1.5 g
15% w/v glycerol	150.0 g

(i)   Dissolve in 1 litre of distilled water.
(ii)   Adjust pH to 5.8 with 0.2 M acetic acid.
(iii)   Sterilize by filtration through a 0.22 μm filter (rinsed with H_2O).

SOB medium

Bactotryptone 2.0% (w/v)	
Yeast extract 0.5% (w/v)	
NaCl	10.0 mM
KCl	2.5 mM
$MgCl_2$	10.0 mM
$MgSO_4$	10.0 mM.

(i)   Combine tryptone, yeast extract, NaCl, KCl, and water.
(ii)   Autoclave for 30−40 min.
(iii)   Add filter-sterilized Mg solutions after cooling.
(iv)   Final pH should be 6.8−7.0.

SOC medium
   Add 20 mM glucose to SOB medium.

Plates
For minimal salts media agar plates:
(i)   Dissolve 20 g of Difco agar in 900 ml of distilled water, autoclave, and add all the solutions separately.
(ii)   Transfer to a 2-litre Erlenmeyer flask, and let the solution cool in a 55−60°C water bath:
(iii)   Pour into standard petri dishes.
For other media add 20 g of Difco agar per litre; for soft agar add 6 g of Difco agar per litre before autoclaving. Melted soft agar can be kept in a 55°C water bath. For M13 plaques, B broth gives a deeper blue colour than YT medium, possibly because of the lack of catabolite repression.

Table 1. (continued)

IPTG solution
   Isopropylthiogalactoside (IPTG) 100 mM
Sterilize by filtration through a 0.22-μm filter (rinsed with H_2O). Store at $-20°C$.

Xgal solution
   5-bromo-4-chloro-3-indolyl-β-D-galactoside (Xgal) 2%
Dissolve in dimethylformamide; wrap in aluminium foil to protect against light. Store at $-20°C$.

Ligase buffer (10×)
   500 mM Tris$-$HCl (pH 7.5)
   100 mM $MgCl_2$
   100 mM dithiothreitol

DNase I buffer
   50 mM Tris$-$HCl (pH 7.5)
   1 mM $MnCl_2$

20× SCP (pH 7.0)
   NaCl   467.5 g
   Na_2HPO_4 340 g
   Na_2EDTA 29.7 g
   HCl conc. 50 ml
Dissolve in 4 litres of H_2O and autoclave.

Nick-translation mix (10×)
   0.2 mM of dATP, dGTP, dTTP (no dCTP)
   500 mM Tris$-$HCl (pH 7.8)
   50 mM $MgCl_2$
   100 mM 2-mercaptoethanol
   0.5 μg BSA (nuclease-free).

TNE buffer
   10 mM Tris$-$HCl (pH 8.0)
   50 mM NaCl
   0.1 mM Na_2EDTA.

Carrier DNA
Dissolve 50 mg of calf thymus DNA (or other source) in 50 ml of TE by shaking in a 37°C incubator overnight.
Store at 4°C.

Hybridization solution (2×)
   SCP (20×) autoclaved   300 ml
   N-laurylsarcosine (25%) (Sigma)   40 ml
   Heparin (1 mg ml^{-1}) (Sigma H7005) 540 mg.
(i)   Dissolve the heparin in 100 ml of H_2O by shaking in a 37°C incubator.
(ii)   Add *N*-laurylsarcosine and SCP, and dilute with autoclaved H_2O to 540 ml.
(iii)   Filter through a 0.45-μm nalgene filter unit.

3.2 **Preparation of DNA**

3.2.1 *M13 RF for cloning*

Since the RF DNA has a lower copy number than the pUC plasmid, a CsCl gradient centrifugation may be useful. A 40-ml culture of infected cells should yield about $20-40$ μg of RF DNA, which is sufficient for most cloning projects. With a toothpick, remove a few infected cells from a single plaque of the M13 strain, add to 2 ml of YT medium (the phage titre will be 10^7-10^8 p.f.u. ml^{-1}, and incubate for ~3 h). Also pick uninfected cells of the host strain from a single colony, add to 10 ml of YT medium, and grow to an $OD_{600} = 0.3$ (<3 h). Add 0.05 ml of the infected to the non-infected culture, and incubate for not longer than 8 h. The titre should be about 2×10^{12} p.f.u. ml^{-1}. This procedure can be scaled down to 1 ml of culture, but yields may be lower and cells may reach a stationary phase at a lower cell density because of a less efficient aeration. Some strains lyse more easily in stationary phase.

(i) Centrifuge infected cells ($8000-10\ 000$ g, 10 min at 4°C), decant the phage supernatant carefully without disturbing the soft pellet, and save for template preparation (see below).

(ii) Suspend the pellet in 4 ml of lysis solution I containing 1 mg ml^{-1} of lysozyme and 10 μg ml^{-1} of RNase. Vortex the pellet to resuspend it thoroughly. Incubate for 5 min in a water bath at 37°C.

(iii) Prepare a solution of 0.2 M NaOH and 1% SDS (lysis solution II) while infected cells are incubating. Add 8 ml of lysis solution II and mix as the solution clears. Incubate for 5 min longer at 37°C to complete lysis.

(iv) Cool the centrifuge tube on ice, add 6 ml of 3 M KOAc, and mix well. Place the tube on ice for 10 min. Pellet the precipitate at 10 000 r.p.m. in a Sorvall SS-34 rotor for 10 min at 4°C.

(v) Filter the supernatant through Miracloth (Calbiochem) into a 30-ml Corex centrifuge tube; avoid collecting any of the white cellular debris. Precipitate RF by adding 0.7 vol of isopropanol. Mix well and leave on ice for 10 min. Pellet the DNA at 10 000 r.p.m. in a Sorvall SS-34 rotor for 10 min at 4°C. Remove the supernatant, vacuum dry the pellet and redissolve it in 4.8 ml of 20 mM Tris−HCl (pH 7.4), add 0.1 ml of 0.5 mg ml^{-1} RNase A, and incubate at 37°C for 30 min. Stop the reaction by adding 4.8 g of CsCl. Before transferring the solution to a Beckman VTi65 ultracentrifuge tube, mix with 0.1 ml of ethidium bromide (10 mg ml^{-1}) and then centrifuge at 60 000 r.p.m. for 16 h at 20°C. To avoid disturbing the gradient during deceleration, turn off the brake at 1000 r.p.m.

(vi) The DNA is observed by UV illumination (300 nm), although a DNA band is usually visible without UV. The denser or lower band is removed by puncturing the tube from the side with a needle and withdrawing the fluid with a syringe. Sometimes a third band of intermediate density, representing single-stranded DNA, can be observed. Only the densest band is retained.

(vii) Extract the recovered CsCl solution with an equal volume of n-butanol to remove the ethidium bromide. Repeat the extraction twice and leave any interface behind.

(viii) Dialyse against 2 litres of distilled water for 6 h. Add 0.1 vol of 3 M NaOAc and 2.5 vol of ethanol, chill at −20°C for 1 h, and pellet the DNA at 12 000 g for 20 min at 4°C in a Sorvall SS-34 rotor. Wash the pellet once with 70%

ethanol, dry under vacuum, and redissolve in 0.2 ml of TE. Remove a 5-μl aliquot and determine the OD_{260} as a measure of the DNA concentration. Adjust the concentration to 100 ng μl^{-1}.

3.2.2 *Phage DNA*

The most critical requirement in 'dideoxy' DNA sequencing is a very clean template; even trace impurities substantially reduce the amount of sequence that can be obtained (Chapter 3, Section 4). Yamamoto *et al.* (34) concentrated phage by low speed centrifugation in the presence of 0.5 M NaCl and 2% PEG 6000. Although higher PEG concentrations increase the yield of M13 phage, other impurities often precipitate as well (35). A high salt content, once added, is not easily removed later by ethanol and results in poor quality sequence. The use of lower concentrations and the replacement of the Cl− by OAc- increases ethanol solubility and gives adequate yields of phage DNA. If the $OD_{260/280}$ ratio is less than 1.75 (indicating a high degree of protein contamination), prepare a new phage culture. As discussed earlier, infected cells in the stationary phase tend to lyse to a slight degree; small RNA fragments would be purified together with phage DNA. To prevent interference by these impurities, an RNase step can be included before phenol extraction.

(i)   Conditions for the growth and precipitation of infected cells are described above. To 9 ml of supernatant add 0.5 ml of 5.0 M NaOAc (pH 7.0) and 0.5 ml of 40% PEG 8000. Mix thoroughly by vortexing and place on ice for 45 min.

(ii)   Collect the phage by centrifugation at 10 000 *g* for 15 min at 4°C, if possible in a swinging-bucket rotor, to avoid streaking the pellet along the centrifuge tube. Discard the supernatant; drain the centrifuge tube upside down on Kimwipes for 20 min and wipe clean using a cotton swab without disturbing the pellet.

(iii)   Suspend the phage pellet in 0.5 ml of 20 mM Tris−HCl (pH 7.5) and transfer to a 1.5-ml microcentrifuge tube. Incubation at 37°C for 30 min in the presence of 5 units of RNase T1 is optional.

(iv)   Deproteinize the phage by adding an equal volume of phenol neutralized with 20 mM Tris−HCl (pH 7.5). Vortex for 2 min, centrifuge, and remove the aqueous phase, leaving any interface behind. Repeat the extraction with an equal volume of a 1:1 phenol-chloroform mixture, and again with an equal volume of chloroform.

(v)   Adjust the volume to 0.47 ml and add 30 μl of 5 M NaOAc (pH 7.0) and 1.25 ml of pure cold ethanol. Precipitate the DNA for at least 3 h at −20°C and collect in a microcentrifuge at full speed for 15 min at 4°C. Clean the tubes with Kimwipes, wash the pellet with 70% ethanol and dry under vacuum.

(vi)   Suspend the dried pellet in 50 μl of Tris−HCl (pH 7.5). Determine the absorption ratio at $OD_{260/280}$; the yield of DNA should be 0.05−0.1 mg.

3.2.3 *Plasmid DNA for cloning*

If templates are prepared from pUC118 and 119 by superinfection with the helper phage M13KO7, DNA is cloned directly in those vectors. In principle, the method used to prepare RF can also be used for pUC plasmids. Inoculate 40 ml of YT containing ampicillin (100 μg ml^{-1}) with a single colony of the strain harbouring the desired pUC plasmid, and grow overnight. The best yields are obtained from stationary cultures, probably because replication of the plasmid continues while cell division stops. Such a culture may yield $0.5 - 1$ mg of plasmid DNA, but usually less with recombinant pUC plasmids. Although it is not necessary to use such large volumes of culture and the additional CsCl gradient centrifugation purification for subcloning, the gradient helps to obtain pure supercoiled plasmid molecules for DNA and in particular cDNA libraries.

Because of the high yields of pUC plasmids, a small scale DNA preparation normally gives sufficient DNA for restriction analysis and subcloning. The DNA is cut by most enzymes, but for some enzymes such as *Kpn*I, it may be necessary first to microdialyse the DNA against a low Tris buffer or the specific restriction buffer.

(i)   Inoculate 2 ml of YT medium containing ampicillin ($100 \ \mu g \ ml^{-1}$) with a single colony and incubate overnight at 37°C.

(ii)   Collect bacteria by centrifuging a culture sample in a 1.5-ml microcentrifuge tube for 2 min. Discard the supernatant by draining the tube for 10 min and removing residual liquid with a cotton swab.

(iii)   Suspend the pellet in 125 μl of lysis solution I containing RNase ($10 \ \mu g \ ml^{-1}$) and lysozyme ($1 \ mg \ ml^{-1}$). Vortex and incubate at 37°C for 5 min.

(iv)   Add 200 μl of stock lysis solution II, mix gently, and place on ice for 10 min.

(v)   Add 150 μl of 3 M KOAc, mix well, and keep on ice for 10 min.

(vi)   Centrifuge, if possible in a swinging-bucket rotor with microcentrifuge tube adaptors, at 10 000 g for 10 min at 4°C; otherwise use a microcentrifuge at full speed for 6 min. Remove the supernatant, without disturbing the white pellet, to a fresh tube.

(vii)   To the supernatant add 280 μl of isopropanol, mix well, place on ice for 10 min, and collect the precipitate by spinning at full speed in a microcentrifuge for 6 min. Wash with 500 μl of 70% ethanol and dry under vacuum.

(viii)   Suspend the pellet in 5 μl of TE and 45 μl of H_2O, which is enough for $15-50$ restriction analyses.

3.2.4 *Plage DNA (normal scale)*

If single-stranded DNA is to be isolated from strains with pUC118 or 119 or their derivatives, superinfection with M13KO7 is necessary. A stock solution of M13KO7 is prepared as described earlier under growth and maintenance. The pUC plasmid must be maintained in the appropriate F′ strain. The presence of the F′ factor is verified by growing cells on M9 minimal medium in the absence of proline. Since the bacterial chromosome carries a large deletion of the *lac* and proline operons, the F′ factor can complement this deficiency.

(i)   Inoculate 10 ml of YT or SOB medium containing ampicillin ($100 \ \mu g \ ml^{-1}$) with a single colony, and grow to $OD_{600} = 0.3$ with moderate aeration, at approximately 200 r.p.m. on a shaker at 37°C (<3 h).

(ii)   Add 100 μl of M13KO7 phage stock with a titre not less than 10^{12} p.f.u. ml^{-1} and replenish the ampicillin to a concentration of 50 $\mu g \ ml^{-1}$. Continue incubation for 60 min.

(iii)   Add 100 μl of a kanamycin ($5 \ mg \ ml^{-1}$) solution, and grow overnight or $14-16$ h with good aeration at approximately $250-300$ r.p.m. on a shaker at 37°C.

(iv)   Collect bacteria at 12 000 *g* for 10 min at 4°C in a Sorval SS-34 rotor and retain the supernatant.
(v)   The supernatant contains the defective phage particles with the single-stranded pUC plasmid (plage) in large excess (at least 100-fold) over the helper phage M13KO7, which can be tested quickly by direct agarose gel electrophoresis (DIGE) (see Section 3.5.1).
(vi)   For template preparation follow the protocol given for M13 single-stranded DNA (see above). The yield of DNA from plage is always lower than from phage particles, but should give at least one fifth of the yield.

3.2.5 *Phage, plage DNA for sequencing (small-scale)*

If many templates are sequenced routinely, it is useful to reduce the culture volume; in this way cells can be collected in microcentrifuge tubes and more samples handled at one time. The growth of host strain and phage infection are as described under RF preparation for phage DNA and in the previous section for superinfection with M13KO7. However, a tube roller should be used for aeration. Even with the smaller quantities, this procedure serves for the T4 progressive-deletion cloning method.

(i)   Fill a microcentrifuge tube with a mature culture and spin at full speed in a microcentrifuge for 6 min. Transfer 1.5 ml of supernatant to a fresh tube without disturbing the pellet.
(ii)   Add 85 μl of 5 M NaOAc (pH 7.0) and 85 μl of 40% PEG 8000, and mix thoroughly by vortexing. Place on ice for 15−30 min.
(iii)   Collect phage in the microcentrifuge at full speed for 6 min. Discard the supernatant; drain the centrifuge tube upside down on Kimwipes for 20 min and wipe clean with a cotton swab without disturbing the phage pellet.
(iv)   Suspend the pellet in 0.3 ml of 20 mM Tris−HCl (pH 7.5). Optionally, incubate at 37°C for 30 min in the presence of 1 unit of RNase T1.
(v)   Deproteinize phage by adding an equal volume of phenol neutralized with 20 mM Tris−HCl (pH 7.5). Vortex for 2 min, centrifuge to separate the phases, and remove the aqueous phase leaving any interface behind. Repeat the extraction with an equal volume of 1:1 phenol-chloroform, and again with an equal volume of chloroform.
(vi)   Adjust the volume to 0.47 ml and add 30 μl of 5 M NaOAc (pH 7.0) and 1.25 ml of pure cold ethanol. Precipitate the DNA at −20°C for at least 3 h and collect by spinning at full speed in a microcentrifuge for 15 min at 4°C. After centrifugation clean the tubes with Kimwipes. Wash the pellet with 70% ethanol and dry under vacuum.
(vii)   Suspend the pellet in 10−20 μl of Tris−HCl (pH 7.5). Determine the amount of DNA by running 1 μl on a gel and comparing it with a standard; the yield should be 5−10 μg.

3.3 Subcloning techniques

The simplest technique for subcloning is to digest the target DNA and to 'shotgun' the resultant fragments into a cloning vector. Several recombinants are then selected from this library and characterized. An analysis of the recognition sites of any single restriction enzyme within a known sequence reveals a problem with their use for subcloning, namely the uneven distribution of the recognition sites. Some locations have a high density of recognition sites, but other regions have none at all. Consequently a recombinant library produced with the aid of a restriction enzyme has inserts varying widely in size. Some will be as small as a few nucleotides and others much larger. This is especially awkward when cloning into M13 vectors, as the small fragments clone at a much higher frequency than larger ones and often the only way to obtain the recombinants containing larger inserts may be to isolate specific fragments for separate cloning. The very large fragments may of course not be present at all in the library, if their size falls outside the effective range of the vector system used.

An independently produced library, using a different enzyme, is then needed to determine the order and orientation of the first set of subfragments and to confirm that each pair of subfragments is truly contiguous with no small fragments between them. Even when it appears that the entire fragment is represented in the subclone library, it is possible that a small fragment has been missed. This error occurs frequently with large projects when, during the mapping phase before cloning, small fragments remain undetected either because their effective concentration as shown by ethidium bromide staining is low in comparison with the larger fragments, or because two or more fragments of similar length migrate together. It may even happen that the small fragments are lost from the bottom of the agarose gel. With shotgun libraries the only way to identify overlapping clones is by trial and error; in practice many restriction-fragment recombinant libraries may have to be screened before a complete set of overlapping clones is obtained (19,20). Apart from missing fragments, another source of error is that it is often assumed that two restriction fragments, cloned within the same recombinant, arise by partial digestion rather than by the equally possible ligation of two non-adjacent pieces from the target DNA. Their actual origin can be determined only by sequencing across the junction regions with template DNA generated with a different enzyme. To overcome these problems, an initial restriction enzyme mapping step serves to identify suitable enzymes which can be used to generate subfragments of the most convenient size and location. However, this can be very time-consuming and complex, especially in the analysis of four base recognition enzymes in genomes the size of several tens of kilobases. If the target DNA is only a few thousand bases long, this problem may be solved simply by using a subcloning strategy to create nested deletions (see Sections 3.3.7−3.3.9).

3.3.1 *Random subcloning*

Random subcloning of DNA fragments of uniform size can be applied to any fragment regardless of size and without prior knowledge of the location of restriction enzyme sites. In addition, overlapping clones can be isolated from a single recombinant library. This is particularly useful for large projects, even whole genomes. Random breaks are introduced into the DNA to generate subfragments, under conditions designed to give subfragments concentrated within a desired size range. The distribution of sizes of subfragments isolated for cloning can be further narrowed by fractionation through agarose gels. The selected subfragments are shotgun-cloned to produce a recombinant library.

The inserts of a random-clone library can arise from any region of the original fragment and can be inserted into the vector in either orientation. Independent isolates from the library must be characterized in terms of their relationship to other clones. Recombinants of subfragments that were consecutive within the original fragment can be identified by cross-hybridization or restriction mapping. With this cloning method for DNA sequencing, it is simpler and quicker in the long run to sequence isolates from the recombinant library at random until overlap between subclones is complete, i.e. until a contiguous double-stranded sequence of the original fragment is produced. By its nature this process produces significant redundancy and overlap of data. Although the effort may seem excessive, it is directed to the sequence reactions, the simplest part of the procedure. High accuracy is assured by data redundancy, which varies with fragment size but compares favourably with the redundancy afforded by other cloning strategies. When the sequence is completed with the aid of only randomly produced clones, each nucleotide is sequenced an average of about five times for a 2-kb fragment and up to eight to ten times for a 25-kb fragment. Irrespective of cloning method, the minimum of redundancy needed for an accurate sequence is that each base is read at least twice on both strands for a total redundancy of four. This is often increased by reading sequence films several times to eliminate typographical errors.

3.3.2 *Self-ligation*

Physical or enzymatic cleavage are used to cause random breaks into uniform subfragments. Both techniques give fairly random results, with some preference for A−T or G−C rich regions. However, the high frequency of fragment ends creates a bias in the accumulating data which can be reduced in two ways. First, the circular plasmid vector with the recombinant fragment can be linearized within the vector region and then enzymatically broken in random fragments (22). After cloning the random fragments into M13, sequences from the plasmid vector and from the recombinant fragment can be screened by plaque hybridization (see below). Second, the target fragment can be removed and self ligated before the breakage step. In this case:

(i)   cut the fragment out of the vector with the appropriate enzyme and purify by agarose gel electrophoresis as described below;

(ii)   dissolve 3 μg of the DNA to be shotgun-cloned in 25 μl of TE. Add 3 μl of 10× ligase buffer and 3 μl of 10 mM ATP;

(iii)   add T4 DNA ligase (2 units for 'sticky end' or 100 units for 'blunt end' ligations) and incubate at 15°C for 2 h.

3.3.3 *DNA sonication*

Breaks can be produced by random shearing simply by forcing a DNA solution through a fine gauge needle. The shear rate depends upon the needle bore, the size and concentration of the DNA, and the force applied to the solution. This method is difficult to reproduce and impractical for most purposes; careful monitoring, usually by agarose gel fractionation, is required to maintain the desired size range. Sonication, on the other hand, is an established method of shearing DNA (10,23−25). The molecules are broken down by high frequency vibrations of a probe inserted in the solution. The shearing action depends upon the power of the sonicator and the length of the DNA molecules. At constant power, the subfragments are reduced in size progressively and predictably.

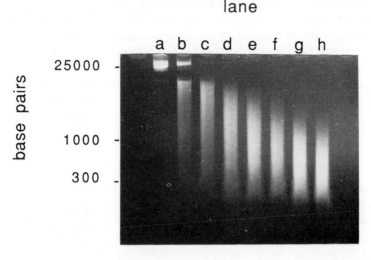

Figure 6. A photograph of a time course of sonication using a Heat Systems Ultrasonics W-375 cup-horn sonicator. A 25-kb recombinant plasmid DNA solution (around 5 μg in 100 μl of water) was sonicated within a Sarstedt microcentrifuge tube clamped 1 mm above the probe, which is submerged in water. Small aliquots were removed at time intervals and run on a 1% agarose minigel (10 cm square) in the presence of ethidium bromide. The time intervals are:- **lane** (**a**) 0 sec, (**b**) 10 sec, (**c**) 20 sec, (**d**) 40 sec, (**e**) 80 sec, (**f**) 160 sec, (**g**) 240 sec and (**h**) 320 sec.

The smaller fragments are more resistant to shear and require longer sonication times. As a consequence, DNA of almost any length or concentration reaches the same molecular weight range at about the same time, with larger fragments breaking down very quickly. This can be seen in *Figure 6*, where each time interval is twice the previous one. Calibrated equipment gives reproducible size distribution in uniform time as a function of power output.

The size of fragment chosen for random cloning into M13 vectors is limited to a few kilobases. This constraint is not imposed by the biology of the phage, which can accommodate larger genomic size simply by an increase in the length of the filament coat. Phage particles of about seven times the unit length have been observed (36). On the other hand, the longer the viral genome, the greater the incidence of deletions, which can cause several problems. Deletions can give rise to double sequences because of the presence of more than one template. They can form colourless plaques which have lost all their insert, and they can produce recombinants missing large pieces and making the interpreted sequence incorrect. Deletion of the primer-annealing site can also occur.

Apart from insert size, some sequences do not clone readily into M13. Whether because of structural considerations or the expression of disadvantageous promoters or gene products, the efficiency of cloning can fall almost to zero. It is often found that when these 'unclonable' regions are reduced to smaller fragments, the cloning efficiencies return to normal. In choosing the minimum size range for cloning into M13, effort should not be wasted. In DNA sequencing, for example, it takes just as much time and effort to obtain a sequence from a very small insert as it does from a very large one, but the return is obviously very different. In this circumstance the minimum size chosen for cloning into the vector should be at least the length of sequence data expected from each recombinant; this is usually 300−600 nucleotides. Nested or unidirectional deletions, as described below, do not avoid the problem of cloning larger fragments. A large insert must be cloned and at least one of the subclones must have the original insert before sequences are deleted from one end. Therefore, the recently developed plage vectors, as described above, should be used with this deletion method, although the preparation of single-stranded DNA always requires superinfection.

3.3.4 *Direct contact probe sonicators*

To avoid superinfection and lower yields of single-stranded DNA, M13 vectors are preferable for the convenience of cloning uniform fragments of less than 1000 bases.

(i)   Place the smallest available probe for the sonicator (a probe larger than a microtip requires excessive sample dilution) into a 0.1% (w/v) solution of sodium dodecylsulphate (SDS) in water and boil for 5−10 min to clean it and remove DNA contamination. Rinse the tip well with water followed by distilled water.

(ii)   Fit the probe to the sonicator, following the manufacturer's instructions.

(iii)   Place the 3 μg of self-ligated DNA into a microcentrifuge tube and dilute to a volume permitting immersion of the probe.

(iv)   Support the tube in an ice bath, with the probe surrounded by solution.

(v)   Sonicate the sample at medium power in bursts of less than 5 sec with periods of about 30 sec between bursts to prevent heating. Normal laboratory equipment gives the size distribution needed for M13 cloning with a total sonication time of approximately 20 sec; this should be determined by calibration as in *Figure 6*.

(vi)   Dismantle and clean the probe as in step (i).

(vii)   Extract the sheared DNA solution once with an equal volume of phenol saturated with TE. Place the upper aqueous phase in a clean tube. Add 0.1 vol of 3 M NaOAc and 2.5 vol of ethanol, chill at −20°C for 1 h, and pellet the DNA at 12 000 *g* for 20 min at 4°C in a Sorvall SS-34 rotor. Wash the pellet once with 70% ethanol, dry under vacuum, and dissolve in 30 μl of ligation buffer.

3.3.5 *Cup-horn sonicators*

Sonicators of the cup-horn type are much easier to use. They have a large probe mounted in a sealed container partially filled with water or another liquid that helps to maintain an even temperature. The sample is sonicated in a vessel immersed in the liquid. This design has two advantages over the direct contact probe. The sample can be kept at a low volume, which eliminates precipitation steps and possible contamination; the probe is never in contact with the DNA solution and need not be cleaned.

(i)   Fill the sonicator cup with distilled water to a depth of about 3 cm above the probe.

(ii)   Place the 3 μg of self-ligated DNA into a microcentrifuge tube similar to the tube used for calibration; the thickness of the tube wall affects the sonication. The DNA need not be precipitated before or after sonication.

(iii)   Fix the tube in a hole in a polystyrene 'raft' resting about 1 mm above the probe.

(iv)   Sonicate the DNA in bursts of less than 60 sec; change the water after each burst to cool the probe. With the Heat Systems Ultrasonics W-375 or similar model, a total sonication time of about 120 sec at maximum power is usually sufficient.

3.3.6 *DNase I*

Random breaks in duplex DNA can be introduced enzymatically using DNase I in the presence of Mn^{2+} (21,22). An agarose gel of the products of a time course of digestion gives results resembling *Figure 6*. In contrast to sonication, however, the effect is linear. Continued digestion produces distributions with progressively smaller mean size, but at a rate directly proportional to incubation time. The response of the DNA sample is affected by temperature, DNA concentration, DNase I concentration, and the age of the DNase I, which loses activity with time. Consequently each experiment requires calibration and is thus less convenient than sonication.

(i)   Dilute 3 μg of self-ligated DNA in 0.1 ml of TE and extract once with an equal volume of phenol saturated with TE. Place the upper aqueous phase in a clean tube. Add 0.1 vol of 3 M NaOAc and 2.5 vol of ethanol, chill at $-20°C$ for 1 h and separate the DNA at 12 000 *g* for 20 min at 4°C in a Sorvall SS-34 rotor. Wash the pellet once with 70% ethanol, dry under vacuum, and dissolve in 90 μl of DNase I buffer.

(ii)   Add 0.25 ng of DNase I (Sigma) from a stock solution made up in 0.15 M NaCl.

(iii) Incubate at 15°C for a time determined by experiment. With freshly prepared stock DNase I, the digestion time is about 10 min.

(iv)   Stop the digestion by adding 1.5 μl of 300 mM Na_2EDTA (pH 8.0). Extract the DNA solution once with an equal volume of phenol saturated with TE. Transfer the upper aqueous phase into the clean tube. Add 0.1 vol of 3 M NaOAc and 2.5 vol of ethanol, chill at $-20°C$ for 1 h, and pellet the DNA at 12 000 *g* for 20 min at 4°C in a Sorvall SS-34 rotor. Wash the pellet once with 70% ethanol, dry under vacuum, and dissolve in 30 μl of ligation buffer.

3.3.7 *Progressive deletions*

Progressive deletion offers an alternative to shotgun sequencing and is usually preferred at the cDNA or single gene level because it provides useful deletion clones for the functional analysis of the sequence. It may be advantageous to carry out the deletion cloning in plage vectors. Since these methods do not produce inserts of uniform size and actually begin with inserts of a few thousand nucleotides, a plage vector may also be required for the resultant increased stability of the high molecular weight single-stranded DNA templates. Progressive deletion reactions also require empirical adjustments, as with the DNase I digestion of DNA. Time courses for both exonuclease III and T4 polymerase must be determined for each enzyme batch. Analysis by agarose gel electrophoresis is preferred, using an appropriate DNA size marker. This provides preliminary information. The sizes of the DNA inserts vary even within time points, and after cloning, recombinant phage or plage must be ordered according to size by DIGE (Section 3.5.1) to provide the least redundancy and the necessary overlapping sequencing information. If this analysis leaves gaps of deletion points, additional deletions can be made, or a few synthetic oligonucleotide primers used to provide the necessary information later.

3.3.8 *ExoIII-ExoVII method*

For this method, it is important to clone the DNA into pUC118 or 119 in such a way that two unique restriction cleavage sites are located in the polylinker between the insert and the universal primer site. Cleavage next to the primer site should generate a single stranded 3'-OH overhang (*Pst*I or *Sst*I), cleavage next to the insert a 3'-OH recessed end or a blunt end. The 3' overhang resists attack by exonuclease III but the 5' overhang or blunt end do not, which provides the basis for unidirectional deletion into the cloned section of the DNA. DNA sequencing can be initiated at various deletion end points. Since exonuclease III requires a divalent cation, the progressive exonuclease reaction can be terminated at various times by the addition of EDTA. In contrast to exonuclease III, the single-strand specific exonuclease VII does not require a divalent cation and all protruding ends can be removed before the recyclization at the blunt ends of the molecule. Therefore, both steps can be carried out in the same buffer without a phenol or ethanol step.

(i)   About 2 μg of the pUC clone containing the fragment is cleaved with the two appropriate restriction endonucleases following the manufacturer's recommendation. Extract restriction enzymes from the aqueous solution with an equal volume of phenol neutralized with 20 mM Tris−HCl (pH 7.5). Vortex for 2 min, centrifuge to separate the phases, and remove the aqueous phase leaving any interface behind. Repeat the extraction first with an equal volume of 1:1 phenol-chloroform, and then with an equal volume of chloroform. Adjust the volume to 94 μl and add 6 μl of 5 M NaOAc (pH 7.0) and 250 μl of pure cold ethanol. Precipitate the DNA at −20°C for at least 3 h and collect it by spinning at full speed in a microcentrifuge for 15 min at 4°C. Clean the tubes with Kimwipes, wash the pellet with 70% ethanol and dry it under vacuum.

(ii)   Suspend the DNA in 40 μl of ExoIII buffer with 8 units of exonuclease III (Boehringer Mannheim), and incubate at 37°C for 20 min. At 60-sec intervals place 2-μl aliquots in an ice-cold tube containing 2 μl of 10 × ExoVII buffer. Start a second tube after 10 min. Each tube then contains the pooled aliquots taken during a 10-min period.

(iii) To each tube add 1 μl of exonuclease VII (Bethesda Research Labs) (100 units ml^{-1}), and incubate for 45 min at 37°C and then for 15 min at 70°C.

(iv)   Proceed with the end repair (Section 3.3.10).

3.3.9 *T4 deletion method*

As an alternative to the double-stranded DNA deletion method described in the previous section, T4 DNA polymerase, which has a single-stranded exonuclease activity specific for the 3'-OH end, can be used to make progressive deletions of a single-stranded DNA. This 3'-OH end has to be generated at the polycloning site, allowing progressive deletions of the cloned DNA from one end. To cleave the ss circular plage or phage DNA in the polylinker, an oligonucleotide is hybridized to this region and the double-stranded DNA cut with the appropriate restriction endonuclease. The oligonucleotide is only partially complementary to the polycloning site. It also contains a T-tail for the pUC and M13 vectors with even numbers, and a C-tail for the ones with odd numbers. All pUC and M13 vectors with even numbers (e.g. pUC118) have a *Hind*III restriction site proximal to the universal sequencing primer; vectors with odd numbers (e.g. pUC119) have an *Eco*RI restriction site proximal to the universal sequencing primer. The tailing reaction depends on which one is used. However, when sequencing the deletion clones, it is preferable to use the oligonucleotide employed here to cleave and then reseal the DNA. This helps to reduce any artefacts in the sequencing reaction arising from the inserted T- or C-tails (*Figure 5*). For the initial template preparation, it may be useful to make a larger quantity (normal scale). After the generation of nested deletions, the small scale method can be resumed.

The inclusion of T4 single-stranded DNA binding protein may facilitate melting of hairpin structures of the single-stranded DNA template, which can otherwise lead to an unequal distribution of nested deletions.

(i) Prepare plage or phage DNA of the clone as described above. Add to 2 μl of single-stranded DNA (1−2 μg), 2 μl of 100 mM Tris−HCl (pH 7.5), 13 μl of H_2O, and 1 μl (40−80 ng, but check supplier's instructions) of oligonucleotide [*Eco*RI (Boehringer Mannheim) for pUC119 and *Hin*dIII (New England Biolabs) for pUC118; for *Eco*RI it should have a sequence like 5′ CGACGGCCAG-TGAATTCCCCCC 3′]. Heat to 65°C for 10 min and cool slowly to 37°C. Add 2 μl of the appropriate enzymes and incubate at 37°C for 1 h.

(ii) In the case of *Eco*RI, heat to 65°C for 10 min; for *Hin*dIII heat to 75−80°C for 15 min to stop the reaction. Use 1 μl of the reaction mixture for agarose gel electrophoresis to check restriction cleavage.

(iii) Keep the remainder on ice, adding 2 μl of 10 × T4 polymerase buffer and 3 μl of a mixture of 0.1 M DTT and 20 mg ml^{-1} BSA. Mix, add 1 μl of T4 DNA polymerase (New England Biolabs) (1 unit μl^{-1}), and place in a water bath at 37°C. Remove 5-μl aliquots at appropriate times, bearing in mind that the polymerase removes about 40 nucleotides per min; heat to 65°C for 5 min to stop the reaction. Remove 1 μl of mixture for agarose gel electrophoresis to check the reaction. Repeat with larger quantities to obtain the necessary nested deletions.

(iv) Pool these reactions and extract enzymes from the aqueous solution with an equal volume of phenol neutralized with 0.5 M Tris−HCl (pH 7.5). Vortex for 2 min, centrifuge, and remove the aqueous phase leaving any interface behind. Repeat the extraction first with an equal volume of 1:1 phenol-chloroform, and then with an equal volume of chloroform. Adjust the volume to 94 μl and add 6 μl of 5 M NaOAc (pH 7.0) and 250 μl of pure cold ethanol. Precipitate the DNA at −20°C for at least 3 h and collect it by spinning at full speed in a microcentrifuge for 15 min at 4°C. Clean the tubes with Kimwipes, wash the pellet with 70% ethanol and dry it under vacuum.

(v) Suspend the DNA in 15 μl of T4 polymerase buffer for terminal deoxynucleotide transferase (TdT) reaction. Add to the solution 2.0 μl of 50 μM dGTP (for *Hin*dIII cleaved DNA, use dATP), 2.0 μl of H_2O, and 0.5 μl of TdT (New England Biolabs) (25 units μl^{-1}) and incubate at 37°C for 20 min. Stop the reaction by heating to 65°C for 10 min. Remove 5 μl for an unligated control.

(vi) To the remaining 20 μl of reaction mixture, add oligonucleotide as in step (i) for resealing the 3′ and 5′ ends. Heat to 65°C for 10 min and cool slowly to 40°C. Add 3 μl of 10 mM ATP, 5 μl of H_2O, and 1 μl of T4 DNA ligase (Boehringer Mannheim) (1−5 units μl^{-1}). Incubate at room temperature for at least 1 h, or as long as overnight. Stop the reaction by adding 3 μl of 200 mM Na_2EDTA. Use 2−5 μl of the ligated DNA for transformation.

3.3.10 *End repair*

Both of the methods described for generating random subfragments, physical shearing and enzymatic digestion, leave fragments with staggered (non-flush) ends. This is also to some extent true for the exoIII – VII progressive deletion. Indeed the shearing process is so severe that damage at the ends is unavoidable. Few, if any, of the subfragments clone directly into vector which has been linearized by means of a suitable restriction enzyme to give flush ends. Before ligation to vector, the subfragments are therefore treated with T4 DNA polymerase and the Klenow fragment of pol I. These enzymes help to produce flush-ended fragments in two ways. The polymerase activity repairs 5' overhangs by extension of the 3' end, and the exonuclease activity removes the 3' overhangs in the same way as exonuclease VII. Since exonuclease III leaves far more extensive 3' overhangs, this procedure includes a special exonuclease VII step.

(i)   For the sheared or DNase I treated DNA, add to the redissolved DNA (30 μl of ligase buffer) 4 μl of 0.5 mM dNTP sequence chase solution (containing all four deoxynucleotides).

(ii)   Add 10 units of T4 DNA polymerase (New England Biolabs) and 10 units of Klenow fragment DNA polymerase (Boehringer Mannheim). Incubate at room temperature for 30 min.

(iii)   Inactivate the enzymes by heating at 70°C for 10 min.

(iv)   Before ligation the repaired DNA from the shearing or DNase I treatment needs further size purification (Section 3.3.11).

(v)   For the samples from the progressive deletion method, add to each reaction 3 μl of 0.1 M MgCl$_2$, 1 μl of large DNA polymerase fragment (Klenow fragment) (Boehringer Mannheim) (500 units ml^{-1}), and 1 μl of an 8 mM solution of deoxyribonucleotides. Incubate at room temperature for 30 min.

(vi)   To a 5-μl aliquot (200 ng of DNA) add 5 μl of 10× ligation buffer, 5 μl of 10 mM ATP, and 2.5 μl of 0.1 M DTT. Adjust the volume to 49 μl with H$_2$O, and add 1 μl of T4 DNA ligase (2000 units ml^{-1}). After incubation at room temperature for 3 h, deproteinize the DNA once with an equal volume of phenol saturated with TE. Place the upper aqueous phase in a clean tube. Add 0.1 vol of 3 M NaOAc and 2.5 vol of ethanol, chill at −20°C for 1 h, and pellet the DNA at 12 000 g for 20 min at 4°C in a Sorvall SS-34 rotor. Wash the pellet once with 70% ethanol, dry it under vacuum, and dissolve it in 40 μl of TE; 10 μl are used for the transformation.

3.3.11 *DNA of uniform size*

Given the size range produced by the sonication or digestion conditions, agarose gel electrophoresis provides the most effective technique for the further selection necessary to remove small fragments. After end repair, the mixture is fractionated through an agarose gel and stained with ethidium bromide. The DNA is eluted from a gel slice containing the desired fraction. The progressive deletion methods also require agarose gel electrophoresis, but after the transformation of E.coli with the ligated DNA samples and only for the purpose of analysing the sizes of recombinant phage by DIGE (see Section 3.5.1). The random cloning techniques require preparative gel electrophoresis before cloning and no further analysis after transformation. The DNA can be removed from the preparative agarose gel in several ways.

3.3.12 *Electroelution*

(i)   Place the slice of gel in a small piece of dialysis tubing, cover with a minimum of electrode buffer, and close both ends.

(ii)   Place it in an electrophoresis apparatus, submerged in electrode buffer, and parallel to the electrodes. This minimizes the length of gel through which the DNA migrates.

(iii)   Electrophorese for a few minutes until the DNA has migrated into the surrounding buffer. When there is a large quantity of DNA, it can be seen without UV light by virtue of the ethidium bromide staining.

(iv)   Reverse the polarity of the electrodes for a few seconds and transfer the buffer from the dialysis tubing to a microcentrifuge tube. Extract the DNA solution once with an equal volume of TE-saturated phenol. Transfer the upper aqueous phase to a clean tube. Add 0.1 vol of 3 M NaOAc and 2.5 vol of ethanol, chill at $-20°C$ for 1 h, and pellet the DNA at 12 000 g for 20 min at 4°C in a Sorvall SS-34 rotor. Wash the pellet once with 70% ethanol, dry it under vacuum, and dissolve it in 50 μl of 0.5 $\times$ TE.

3.3.13 *Phenol extraction*

If low gelling temperature agarose is used, the slice can be melted at about 67°C and the agarose removed from the aqueous DNA solution by several phenol extractions followed by ethanol precipitation.

(i)   Place the excised gel slice in a 1.5-ml microcentrifuge tube and melt it at 67°C.

(ii)   Quickly add an equal volume of TE-saturated phenol and vortex thoroughly. Extract for 10 min with occasional vortexing.

(iii)   Spin in a microcentrifuge for 2 min and transfer the upper aqueous phase, free of any interface material, to a clean tube.

(iv)   Repeat the extraction three or four times until no interface material can be seen and the two phases separate quickly.

(v)   Add 0.1 vol of 3 M NaOAc and 2.5 vol of ethanol, chill at $-20°C$ for 1 h, and pellet the DNA at 12 000 g for 20 min at 4°C in a Sorvall SS-34 rotor. Wash the pellet once with 70% ethanol, dry it under vacuum, and dissolve it in 50 μl of 0.5 $\times$ TE.

3.3.14 *QN (quaternary ammonium) butanol extraction*

Phenol extraction does not always completely remove the enzyme inhibitors associated with some batches of agarose. Other solvents can be used to help minimize this inhibition of enzyme activity and transformation.

(i)   Place the gel slice in a siliconized microcentrifuge tube and melt the agarose at 65°C. Estimate the volume of molten solution and place in a water bath at 37°C. Add equal volumes of QN butanol and QN aqueous solution. Mix well and spin at full speed in a microcentrifuge for 4 min.

(ii)   Remove the upper butanol phase and transfer to a fresh tube. Repeat the extraction from the aqueous solution first with an equal volume of QN butanol and then with QN butanol and QN aqueous solution as in the above step.

(iii)   Remove the upper butanol phase again. Add one-fourth volume of 0.1 M NaOAc to the combined butanol phases, mix well, and separate the phases in a microcentrifuge as described above. Transfer the aqueous phase to a fresh tube and repeat the butanol extraction with this aqueous phase.

(iv)   Very slowly add an equal volume of chloroform to the combined aqueous phases and place on ice to precipitate the Ctab. Centrifuge for 2 min and transfer the supernatant to a fresh tube. Residual chloroform can be extracted with ether and evaporated in a gentle stream of air. Precipitate DNA by adding 2.5 vol of cold ethanol. Keep at −20°C for 3 h and centrifuge to pellet the DNA.

3.3.15 *Extraction with Spin-X tubes*

Short DNA molecules are readily eluted from agarose in Spin-X tubes (Costar), which are standard 2-ml microcentrifuge tubes containing a removable insert with a filtration membrane base. On centrifugation of the gel slice placed in the insert, the aqueous portion of the gel containing the DNA passes through the membrane into the tube, leaving dry agarose particles in the insert.

(i)   Place the microcentrifuge tube containing the gel slice in an isopropanol bath with dry ice for 10 min to break down the gel structure.

(ii)   Transfer the gel slice to the insert of a Spin-X tube and spin in a microcentrifuge for 10−15 min, until only dry material remains in the insert. (Microcentrifuges with horizontal rotors, such as the Eppendorf 5413, use the full membrane surface and are advantageous for this application).

(iii)   Discard the insert. Extract the DNA solution once with an equal volume of TE-saturated phenol. Transfer the upper aqueous phase to a clean tube. Add 0.1 vol of 3 M NaOAc and 2.5 vol of ethanol, chill at −20°C for 1 h, and pellet the DNA at 12 000 g for 20 min at 4°C in a Sorvall SS-34 rotor. Wash the pellet with 70% ethanol, dry it under vacuum, and dissolve it in 50 μl of 0.5 × TE.

3.3.16 *Ligation of DNA of uniform size*

DNA eluted or extracted from the agarose is now ready for ligation. The fragment solution can be used in standard ligation experiments with linearized, blunt ended vector (phosphatased if required).

(i)   To a 5-μl (200 ng of DNA) aliquot add 5 μl of 10 × ligation buffer, 5 μl of 10 mM ATP, and 2.5 μl of 0.1 M DTT.

(ii)   Adjust the volume to 49 μl with H_2O, and add 1 μl of T4 DNA ligase (Boehringer Mannheim) (2000 units ml^{-1}).

(iii)   After incubation at room temperature for 3 h, add 2 μl of carrier DNA, and deproteinize once with an equal volume of TE-saturated phenol. Remove the upper aqueous phase to a clean tube. Add 0.1 vol of 3 M NaOAc and 2.5 vol of ethanol, chill at −20°C for 1 h, and pellet the DNA at 12 000 g for 20 min at 4°C in a Sorvall SS-34 rotor. Wash the pellet once with 70% ethanol, dry it under vacuum, and redissolve it in 40 μl of TE; 10 μl are used for transformation.

3.4 **Transformation**

3.4.1 *High efficiency*

Using the methods described above enough ligated DNA is produced for subcloning by the normal transformation procedure. However, in some shotgun-cloning experiments, it is desirable to obtain a large phage library. In particular when starting from a λ clone, the recovered quantities of DNA may be small and a highly efficient transformation procedure, such as ordinarily needed for a cDNA library, may be helpful. The following protocol is adapted from the procedure described by D.Hanahan (37), using the strain DH5.

(i)   Inoculate 10 ml of SOB medium with a single colony of the host strain. Glassware must be scrupulously clean; detergent residues create a major problem. Autoclave with pure water and add medium afterwards.

(ii)   Incubate the culture at 37°C with moderate agitation (200−250 r.p.m.) until the cell density reaches $4-7 \times 10^7$ cells per ml, which is about $OD_{550} = 0.6$ for DH5. For growth-curve determination, see Hackett *et al.* (33). Collect the culture in cold 50-ml Falcon tubes, and chill on ice for 10−15 min.

(iii)   Centrifuge the cells at 750−1000 g (2000−3000 r.p.m. in a clinical centrifuge) for 15 min at 4°C. Drain the tubes thoroughly and suspend the cells in one-third of the culture volume of TFB by mild vortexing. Keep on ice for 10−15 min. Repeat the centrifugation and drain the tubes carefully. Suspend the cells in eight-hundredths of the culture volume of TFB.

(iv)   Add 3.5% by vol of DMSO and DTT solution (DnD). Inject the DnD into the centre of the cell suspension and mix gently. Keep the mixture on ice for 10−15 min longer. Add a second aliquot of DnD to the cells, adjusting to a final concentration of 7%. Keep on ice for an additional 10−20 min.

(v)   Transfer 210-μl aliquots from the cell suspension into cold Falcon (2059) tubes. Add DNA samples (<20 μl) to each tube. Swirl gently to mix and keep on ice for 20−40 min.

(vi)   Heat shock the transformation mixtures by insertion into a water bath at 42°C for 90 sec. Immediately chill the tubes on ice for 2−4 min.

(vii)   Add 800 μl of SOC to each tube, and incubate at 37°C for 30−60 min with moderate agitation (200 r.p.m.).

(viii) Plate on SOB plates with appropriate selection, or mix with soft agar, host cells, Xgal, and IPTG for generating M13 plaques.

3.4.2 *Small scale*

If the target DNA for subcloning can easily be obtained in larger quantities, the yield of transformation from ligated DNA is less critical and the above procedure can be simplified.

(i)   Streak the host strain on an SOB plate to produce single colonies and grow overnight. Gently disperse two or three colonies by vortexing in 200 μl of TFB. When colonies are transferred to such a small volume, avoid agar contamination which inhibits transformation; keep on ice for 15−30 min.

(ii)   Inject 7 μl of DnD into the centre of the cell suspension, swirl the tube to mix, and keep on ice for 15−20 min. Repeat the addition of DnD exactly as before, and again keep on ice for 15−20 min.

(iii)   Add DNA samples (<20 μl) to each tube. Swirl gently to mix, and keep on ice for 20−40 min.

(iv)   Heat shock the transformation mixtures by insertion into a water bath at 42°C for 90 sec. Immediately chill the tubes on ice for 2−4 min.

(v)   Add 800 μl of SOC to each tube, and incubate with moderate agitation (200 r.p.m.) at 37°C for 30−60 min.

(vi)   Plate on SOB plates with appropriate selection, or mix with soft agar, host cells, Xgal, and IPTG for generating M13 plaques.

3.4.3 *Competent cells for future use*

It may sometimes be useful to have competent cells ready for transformation with a ligation mixture. They can be prepared and stored for up to 3 months without loss of transformation efficiency. For a number of common host strains, such cells are now also commercially available.

(i)   With a single colony from the strain used as host, inoculate 10 ml of SOB medium. Combine 10 inocula in a 1 litre flask. Glassware must be scrupulously clean; detergent residues create a major problem. Autoclave with pure water and add medium afterwards.

(ii)   Incubate the culture at 37°C with moderate agitation (200−250 r.p.m.) until the cell density reaches $4−7 \times 10^7$ cells per ml which is about $OD_{550} = 0.6$ for DH5. For growth-curve determination, see Hackett *et al.* (33). Collect the culture in cold 50-ml Falcon tubes and chill on ice for 10−15 min.

(iii) Centrifuge the cells at 750−1000 g (2000−3000 r.p.m. in a clinical centrifuge) for 15 min at 4°C. Drain the tubes thoroughly and suspend the cells in one-third of the culture volume of TFB by mild vortexing. Keep on ice for 10−15 min. Repeat the centrifugation and drain the tubes carefully. Suspend the cells in eight-hundredths of the culture volume of RF1.

(iv)   Transfer 0.5-ml aliquots from the cell suspension into cold Eppendorf tubes, and immediately freeze the samples in a dry ice-ethanol bath. Store the frozen cells at −70°C until used.

(v)   Thaw the frozen cells at room temperature until the ice just disappears. Transfer 100-μl aliquots from the cell suspension into cold Falcon (#2059) tubes. Add DNA samples (<10 μl) to each tube. Swirl gently to mix and keep on ice for 10−60 min; 20 min is close to optimal.

(vi)   Heat shock the transformation mixtures by immersion in a water bath at 42°C for 90 sec. Immediately chill the tubes on ice for 2−4 min.

(vii) Add 400 μl of SOC to each tube, and incubate with moderate agitation (200 r.p.m.) at 37°C for 30−60 min.

(viii) Plate 100 μl on fresh SOB plates with appropriate selection, or mix with soft agar, host cells, Xgal, and IPTG for generating M13 plaques.

3.5 Screening of recombinants

3.5.1 *DIGE*

Analysis by size is a critical step for recombinant phage or plage from nested deletions. If DNA fragments of more than 200 base pairs are cloned into M13, insertions are large enough to be detected by direct agarose gel electrophoresis of DNA from cultures disrupted by SDS. This discrimination is even clearer with plage vectors, since the vector itself is about 2.5 times smaller and the insert larger in proportion to the vector. As a matter of principle, use the small phage or plage growth volume and initiate the cultures from single colourless plaques (phage) or colourless colonies (plage). For plage, superinfection with M13KO7 is necessary. Follow the same steps until cultures are cleared of bacteria by centrifugation in 1.5-ml microcentrifuge tubes.

(i)   Withdraw 20 μl of supernatant and mix with 1 μl of 2% SDS and 3 μl of loading buffer.
(ii)   Electrophorese overnight through a 0.7% agarose gel in Tris−borate buffer at 100 V (0.7−0.8 V cm^{-1}). Use known recombinant phage as size marker.
(iii)   Stain with ethidium bromide and photograph in UV light as for DNA restriction analysis.

Although it is hard to detect a difference of 200 base pairs, it is easy to choose the right size of recombinant phage from the shotgun-cloning experiment. For progressive deletions another run is necessary, in which the recombinants are arranged by increasing size between vector on one side and full length recombinant on the other side of the gel. In the second run, adjustments should be made in loading equal amounts of phage, to prevent overloading of lanes which would cause the DNA to run abnormally.

3.5.2 *C-test*

This protocol is a very simple means of determining the orientation of the inserts cloned into phage or plage DNA. It is particularly useful for orienting shotgun clones (19). The single-stranded DNA of a larger insert with established orientation is used in a hybridization with the phage DNA of an unknown subclone. If the two single-stranded DNAs have complementary inserts, they hybridize across them and have altered mobility in the DIGE. Grow phage or plage as described for the DIGE and clear the cultures by centrifugation in Eppendorf tubes.

(i)   Combine 20 μl of supernatant from each sample in a fresh tube with 1 μl of 2% SDS and 4 μl of 20 × SSC. Cover with light mineral oil and incubate at 65°C for 1 h.
(ii)   Remove the aqueous phase under the oil, add 5 μl of 10 × loading buffer, and electrophorese at 100 V (0.7−0.8 V cm^{-1}) overnight through a 0.7% agarose gel in Tris−borate buffer. Use known recombinant phage with known complementary inserts and the two samples by themselves as markers.
(iii)   Stain with ethidium bromide and photograph in UV light as for DNA restriction analysis.

The hybridization never goes to completion. Therefore, a positive result should always have three bands, corresponding to the two free DNAs and the figure-of-eight structure, which has lowest mobility.

3.5.3 *Phage hybridization*

Because several thousand phage particles are produced per cell, the phage DNA is highly enriched and is ready for hybridization experiments as single-stranded DNA. A phage library can readily be screened by lifting plaques directly from the plate onto nitrocellulose filters. This is necessary, for example, if the entire recombinant plasmid DNA is cloned by the shotgun procedure. The aim is not to sequence the plasmid but rather the insert. With the aid of the pure plasmid as a hybridization probe, recombinant phage can be detected that contain plasmid DNA. Mutants can similarly be screened from an oligonucleotide site-directed mutagenesis experiment; the oligonucleotide is used as a probe under increasingly stringent conditions (38).

(i)   The soft agar used to plate the phage should be hardened for 1 h in a refrigerator.

(ii)   Place a nitrocellulose filter on the soft agar with the plaques. Contact should be tight, leaving no air bubbles. The filter should darken evenly by taking up moisture. The absorption of the blue colour from vector phage should be visible.

(iii)   After 5 min, lift the filter from the agar surface and air dry it on a Whatman 3MM filter. Bake for 2 h at 80°C under vacuum.

(iv)   As a probe use about 500 ng of plasmid DNA. Take 5 μl of DNA from a stock solution (100 μg ml^{-1} in TE) and add to 5 μl of 10 × nick-translation mix, 10 μl of [α-^{32}P]dCTP (3000 Ci mmol^{-1}, 156 pmol per 50 μl reaction, final concentration 3.12 μM), and 25 μl of H$_2$O. Mix and add 5 μl of DNaseI/DNA polymerase mixture as recommended by the manufacturer (e.g. Bethesda Research Labs). Incubate at 15°C for 60 min. The reaction is terminated by the addition of 5 μl of 300 mM Na$_2$EDTA (pH 8.0).

(v)   Prepare a small Sephadex G-50 (medium) column (0.9 × 15 cm) equilibrated with TNE buffer. Load the solution containing the nick-translated plasmid onto the column and elute with TNE buffer. Discard surplus dCTP in a suitable waste container and save the labelled DNA fraction, which is the first peak of activity to be eluted. Adjust the volume to 375 μl with TNE buffer, add 1 μl of carrier DNA, 24 μl of 5 M NaOAc (pH 7.0), and 1 ml of cold ethanol. Precipitate the DNA at −20°C for at least 3 h and collect it by spinning at full speed in a microcentrifuge for 15 min at 4°C. Clean the tubes with Kimwipes, wash the pellet with 70% ethanol and dry it under vacuum.

(vi)   Place the blot sample in a hybridization bag and add 20 ml of hybridization solution. Remove air bubbles and heat-seal the bag. Place it in an oven at 65°C for 15 min.

(vii)   Boil the nick-translated probe in approximately 2 ml of water and 200 μl carrier DNA for 5 min. Place the tube on ice. Add 2 ml of 2 × hybridization solution and mix.

(viii)   Cut open the sealed bag, add the probe, and reseal; remove bubbles. Place in an oven overnight at 65°C. The radioactivity of solutions in the bag may be checked by a minimonitor.

(ix)   Cut the bag open and remove the probing solution. This may be stored at −20°C and reused two or three times after boiling for five minutes prior to use.

(x)   Wash the blot twice with 2× SCP−1% SDS for 15 min in a square casserole dish in a shaking water bath at 65°C. For a more thorough wash, dilute the 2 × SCP to 0.1−0.2× and wash at 65°C for an additional 15 min.

(xi)   Blot dry the filter and wrap with Saran. Expose overnight in a freezer at −70°C to XAR-5 film (Kodak) with one or two intensifying screen(s). To remove hybridized probe from a blot prior to reprobing, place it in a boiling 0.2 × SCP−0.1% SDS wash for around 10 min with gentle shaking. Check the blot with a minimonitor. If it is still radioactive, wash it again.

(xii)   For a strand-specific probe, use M13 probe primer (10,17). Add to 1 μl of primer

(~ 0.3 ng) 1 μl of template (50 ng), 5 μl of 10 × nick-translation mixture, and 28 μl of H_2O. Heat to 55°C for 5 min, and cool to room temperature. Add 10 μl of [α-^{32}P]dCTP (6000 Ci mmol^{-1}, 156 pmol per 50 μl reaction, final concentration 3.12 μM) and 5 μl of DNA polymerase (Klenow fragment, Boehringer Mannheim, 0.5 units). DNA synthesis proceeds at 15°C for 60 min. Terminate the reaction by the addition of 5 μl of 300 mM Na_2 EDTA (pH 8.0).

(xiii) Prepare a small Sephadex G-50 (medium) column (0.9 × 15 cm) equilibrated with TNE buffer. Load the solution containing the labelled DNA onto the column and elute with TNE buffer. Discard surplus dCTP in suitable waste container and save the labelled DNA fraction. Adjust the volume to 375 μl with TNE buffer, add 1 μl of carrier DNA, 24 μl of 5 M NaOAc (pH 7.0), and 1 ml of cold ethanol. Precipitate the DNA at -20°C for at least 3 h and collect it by spinning at full speed in a microcentrifuge for 15 min at 4°C. Clean the tubes with Kimwipes, wash the pellet with 70% ethanol and dry it under vacuum.

(xiv) For hybridization, resuspend DNA as described above, but do not boil the probe. This would result in melting of the labelled DNA from the single-stranded probe.

4. ACKNOWLEDGEMENTS

We thank K.Elliston and J.Vieira for their input into various procedures and illustrations. J.M. was supported in part by grant DE-FG05-85ER13367 from the US Department of Energy.

5. REFERENCES

1. Messing,J., Gronenborn,B., Muller-Hill,B. and Hofschneider,P.H. (1977) *Proc. Natl. Acad. Sci. USA,* **74**, 3642.
2. Achtman,M., Willets,H. and Clark,A.J. (1971) *J. Bacteriol.,* **106**, 529.
3. Messing,J. (1979) In *Recombinant DNA Technical Bulletin.* NIH Publication No. 79-99,2, No. 2. 43.
4. Dotto,G.P., Horiuchi,K. and Zinder,N.D. (1982) *Proc. Natl. Acad. Sci. USA,* **79**, 7122.
5. Staudenbauer,W.L., Kessler-Liebscher,B.E., Schneck,P.K., van Dorp,B. and Hofschneider,P.H. (1978) The *The Single-Stranded DNA Phages.* Denhardt,D.T., Dressler,D. and Ray,D.S. (eds), Cold Spring Harbor Laboratory Press, New York, p. 369.
6. Dotto,G.P. and Zinder,N.D. (1984) *Nature,* **311**, 279.
7. Landy,A. Olchowski,E. and Ross,W. (1974) *Mol. Gen. Genet.,* **133**, 273.
8. Langley,K.E., Villarejo,M.R. Fowler,A.V., Zamenhof,P.J. and Zabin,I. (1975) *Proc. Natl. Acad. Sci. USA,* **72**, 1254.
9. Calos,M. (1978) *Nature,* **274**, 762.
10. Messing,J. (1983) In *Methods in Enzymology.* Wu,R., Grossman,L., Moldave,K. (eds), Academic Press Inc., London and New York, Vol. 101, p. 20.
11. Gronenborn,B. and Messing,J. (1978) *Nature,* **272**, 375.
12. Vieira,J. and Messing,J. (1982) *Gene,* **19**, 259.
13. Messing,J., Vieira,J. and Gardner,R.C. (1982) In *In vitro mutagenesis.* Schleif,R., Shortle,D. and Wallace,B. (eds), Cold Spring Harbor Laboratory Press, New York, p. 52.
14. Messing,J. and Vieira,J. (1982) *Gene,* **19**, 269.
15. Yanisch-Perron,C., Vieira,J. and Messing,J. (1985) *Gene,* **33**, 103.
16. Heidecker,G., Messing,J. and Gronenborn,B. (1980) *Gene,* **10**, 69.
17. Hu,N.-T. and Messing,J. (1982) *Gene,* **17**, 271.
18. Messing,J., Crea,R. and Seeburg,P.H. (1981) *Nucleic Acids Res.,* **9**, 309.
19. Gardner,R.C., Howarth,A.J., Hahn,P., Brown-Leudi,M., Shepherd,R.J. and Messing,J. (1981) *Nucleic Acids Res.,* **9**, 2871.
20. Sanger,F., Coulson,A.R., Hong,G.F., Hill,D.F. and Peterson,G.B. (1982) *J. Mol. Biol.,* **162**, 729.
21. Anderson,S. (1981) *Nucleic Acids Res.,* **9**, 3015.
22. Messing,J. and Seeburg,P.H. (1981) In *Developmental Biology using purified genes, ICN-UCLA Symposia on Molecular and Cellular Biology.* Brown,D. and Fox,F. (eds), Academic Press, New York, Vol. XXIII, p. 659.

23. Deininger,P.L. (1983) *Anal. Biochem.*, **129**, 216.
24. Bankier,A.T. and Barrell,B.G. (1983) In *Techniques in Life Sciences*. Flavell,R.A. (ed.), Elsevier Scientific Publishers, Ireland, Ltd. Vol. B508, p. 1.
25. Bankier,A.T., Weston,K.M. and Barrell,B.G. (1987) In *Methods in Enzymology*. Wu,R. (ed.), Academic Press Inc., London and New York, Vol. 155, p. 51.
26. Poncz,M., Solowiejczyk,D., Ballantine,M., Schwartz,E. and Surrey,S. (1982) *Proc. Natl. Acad. Sci. USA*, **79**, 4298.
27. Guo,L.-H. and Wu,R. (1982) *Nucleic Acids Res.*, **10**, 2065.
28. Dale,R.M.K., McClure,B.A. and Houchins,J.P. (1985) *Plasmid*, **13**, 31.
29. Cleary,J.M. and Ray,D.S. (1980) *Proc. Natl. Acad. Sci. USA*, **77**, 4638.
30. Griffith,J. and Kornberg,A. (1974) *Virology*, **59**, 139.
31. Dente,L., Cesareni,G. and Cortese,R. (1983) *Nucleic Acids Res.*, **11**, 1645.
32. Vieira,J. and Messing,J. (1987) In *Methods in Enzymology*. Wu,R. and Grossman,L. (eds), Academic Press Inc., London and New York, Vol 153, p.3.
33. Hackett,P.B., Fuchs,J.A. and Messing,J.W. (1988) In *An Introduction to Recombinant DNA Techniques. Basic Experiments in Gene Manipulations*. 2nd edn, Benjamin Cummings Publishing Co., Menlo Park, CA.
34. Yamamoto,K.R., Alberts,B.M., Benzinger,R., Lawhorne,L. and Treiber,G. (1970) *Virology*, **40**, 734.
35. Carlson,J. and Messing,J. (1984) *J. Biotech.*, **1**, 253.
36. Messing,J. (1981) In *Recombinant DNA, Proc. of the Third Cleveland Symp. on Macromolecules*. Walton,A.G. (ed.), Elsevier Scientific Publishers, Amsterdam, Netherlands, p. 143.
37. Hanahan,D. (1983) *J. Mol. Biol.*, **166**, 557.
38. Norrander,J., Kempe,T. and Messing,J. (1983) *Gene*, **26**, 101.

CHAPTER 2

Sequencing single-stranded DNA using the chain-termination method

A.T.BANKIER and B.G.BARRELL

1. INTRODUCTION

1.1 DNA sequencing by primed synthesis

In the previous chapter the preparation of recombinant clones using M13 or M13 derived vectors for sequence analysis by the chain-termination method (1) was described. This chapter is concerned with procedures for preparing the single-stranded template DNA, their sequence analysis using 96-well microtitre trays (2) and the subsequent fractionation of the reaction products on linear (3), gradient (4) or wedge (5) polyacrylamide gels. The automation of the sequence reactions using a Beckman Biomek 1000 automatic pipettor is also described.

The principle of the chain-termination method is simple. A primer, typically a synthetic oligonucleotide 17 bases in length, is annealed to its complementary sequence on a single-stranded DNA template. Ideally the position of hybridization should be $20-30$ nucleotides to the 3' side of the template sequence to be determined. This is usually a position just outside the vector cloning region, so that a single 'universal' primer can be used for all inserts. The primer/template duplex is then used as a substrate for chain extension from the 3' end of the primer by Klenow fragment DNA polymerase I, which copies the template by synthesizing the complementary sequence using deoxynucleoside 5' triphosphates (dNTPs) as precursors. The Klenow fragment, derived from *Escherichia coli* DNA pol I, is used because it lacks the $5'-3'$ exonuclease activity of the naturally occurring enzyme. Other polymerases such as reverse transcriptase, Taq polymerase and bacteriophage T7 DNA polymerase can also be used in sequence reactions. Four separate synthesis reactions are carried out, each with a small amount of a different dideoxynucleotide (ddNTP) present. Incorporation of ddNTPs causes chain-termination since they lack a 3'-hydroxyl group. Random low level incorporation of a specific ddNTP, in competition with the normal dNTP analogue, will result in a mixture of different length chains, all starting at the 5' end of the primer, and ending at every possible position where the specific ddNTP can be incorporated in place of the dNTP. The average length of the resulting chains can be altered by changing the ddNTP/dNTP ratio within the reaction mixture. For example, increasing the relative concentration of the ddNTP will result in a shorter average chain length, since the frequency of its incorporation will increase. With very little experimentation a ratio of ddNTP/dNTP can easily be chosen which can be used for any length of sequence determination, from a few bases up to the maximum which can be resolved on polyacrylamide gels. This means that pre-mixed nucleotide solutions can be used, greatly simplifying the procedure.

In the second step of the sequence reactions, any chains which are not synthesis blocked by the incorporation of a ddNTP are chased out to higher molecular weight lengths by a further incubation in the presence of increased concentrations of dNTPs. This ensures that all the synthesized molecules, within the resolvable size range, are terminated with a specific ddNTP. These four sets of products are then fractionated, alongside each other, on a polyacrylamide gel.

Substituting a radioactively labelled analogue for one of the dNTPs in the reactions provides a means of visualizing the separated bands by autoradiography, using standard X-ray film. Separation on polyacrylamide gels is according to chain length, and reading the sequence on a photographic image of the gel is a matter of finding which track has the shortest fragment, which track has the next shortest fragment, and so on up the gel, as far as the resolution allows. Around 300 bases can be read off a single loading run on a 50-cm long buffer gradient gel as described in Section 5.7.3. Longer sequences, up to a practical limit of around 500 bases, can be obtained by running the same sample for a longer electrophoresis time and combining the data from both the long and short runs.

An alternative labelling technique employs either an oligonucleotide primer or dideoxy nucleotide analogues which have a fluorescent group attached (6,7 and Chapter 8). All the sequence reaction conditions are essentially the same as described for radioactive sequencing. The reaction products, however, are detected directly within the gel, during the run, by excitation of these groups with a laser, and detecting the emitted fluorescence at a fixed vertical position in the gel. The passing bands are detected and analysed whilst electrophoresis is performed. This has the advantage of obtaining, in 'real' time, the maximum possible data from a single run on a short gel. The passing of the bands can be continuously monitored until the resolution is insufficient to ascribe unambiguously an order to the signals. The bands which are electrophoresed off the bottom of the gel have already been processed and are no longer required.

The current horizontal resolution of the detectors used limits the total number of tracks which can be monitored, and so the capacity of these systems is somewhat limited. This has been partially compensated for by the ability to use a different wavelength fluorescence specifically for each terminator reaction. All four reaction sets can be co-electrophoresed in the same track, the colour of the fluorescence being used to differentiate between the products. Even using this elegant approach the currently available machines are limited to 16 templates per gel and therefore per machine.

The methods described below are applicable to sequencing any recombinant M13 DNA with an appropriate primer. If specific primers are being made on an automatic synthesizer then these can normally be used unpurified after dilution in TE buffer or water. If large numbers of clones are to be sequenced, for example by the shotgun approach (Chapter 1 and Section 1.2 below), then with experience, 48 templates per day has been found to be a convenient number. This requires four gel apparatuses each capable of fractionating the sequencing reactions for 12 clones and four power supplies or a similar appropriate configuration. The sequence reactions can be carried out and the gels can be poured and loaded in the morning. In the afternoon, after electrophoresis, the gels are dried, autoradiographed overnight and the films read the next day.

1.2 **Choosing a sequencing strategy**

In a few situations, the limit to the amount of data obtainable from each sequence reaction (around 500 bp) is adequate, for example when checking inserts for specific mutations. More often this will only be a small part of the sequencing goal. If the DNA to be sequenced is larger than this limit in size, cloning strategies as described in Chapter 1 become a major consideration in planning a sequencing project. The route chosen will depend upon the size of the sequencing project, the availability of specialist equipment, the ultimate aims of the project and practical experience and familiarity with the techniques employed. The most significant factor is the size of DNA fragment to be sequenced. Each of the cloning strategies has its own benefits and pitfalls which assume greater or lesser significance depending upon the size of the task.

Where the sequence to be determined is of a length around 1 kb, subcloning using restriction enzymes (Section 1.2.1) or sequence walking by using specific oligonucleotide primers (Sections 1.2.2 and 1.2.3) are the most effective. These include various techniques of generating ordered deletions of the inserted fragment (Section 1.2.4) and random subcloning (Section 1.2.5). When the size of the project exceeds 3 kb or many different fragments are to be sequenced, random shotgun-cloning is by far the most effective approach.

1.2.1 *Cloning using restriction enzymes*

The polylinker region of the M13 series of vectors, and the plasmids derived from them, can be used to clone a very wide variety of fragments generated using restriction enzymes. The subfragments generated by the enzymatic digestion can be isolated and cloned specifically or, more commonly, the digestion products can be shotgun-cloned (as shown in *Figure 1*) and the independent recombinants purified and analysed.

Advantages:

(i)   The cloning step is very fast and easy.

Disadvantages:

(i)   Smaller fragments clone more readily into M13 vectors.
(ii)   The larger fragments are often not represented in the library.
(iii)   The larger inserts cannot be sequenced across their full length without further subcloning.
(iv)   Several independent libraries are needed for larger projects.
(v)   More than one subfragment can be cloned within the same recombinant.
(vi)   Determining the small sequences makes inefficient use of gel space.

1.2.2 *Primer directed sequencing*

If the entire fragment to be sequenced can be cloned into an M13 vector, the sequence can be determined across its full length using different oligonucleotide primers. The first sequence can be obtained using the universal primer. The sequence gained from this priming and each subsequent one can be used to choose a suitable sequence for the next primer. This is chosen to be close to the limit of the currently determined

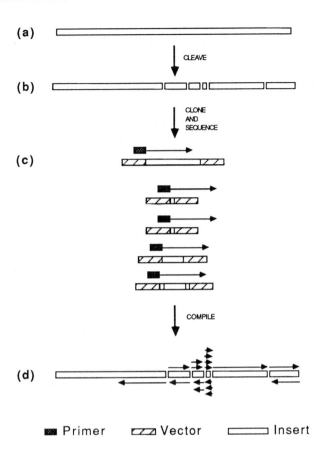

Figure 1. Subcloning/sequencing using restriction enzymes. The fragment to be sequenced in (**a**), is cleaved with a restriction enzyme (**b**), cloned into M13 and sequenced using a universal primer (**c**) and the data compiled (**d**).

sequence and to have limited homology to the vector sequence. This cycle of sequence, oligonucleotide synthesis, sequence can be continued until the end of the insert is reached (see *Figure 2*).

Advantages:

(i)   The method has the maximum possible efficiency since very little redundant data is created.

(ii)   A single template preparation can be used for all the sequence reactions.

(iii)   Getting the sequence of the second strand is quick since all the primers can be predicted from the first strand sequence.

Disadvantages:

(i)   It has to be possible to clone the entire piece of DNA in both orientations to get both strands.

(ii)   The first strand sequence is gained slowly since each step can be taken only one at a time.

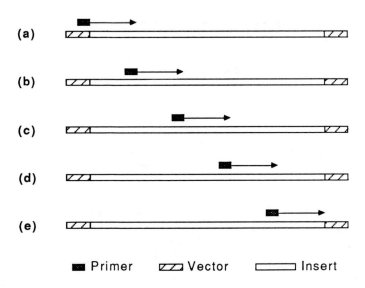

Figure 2. Sequence determination using specific oligonucleotide primers. A universal primer is used to determine the sequence, as far as possible, into the insert, as depicted by the arrow in (**a**). The sequence thus gained is used to select a second, specific primer which is then used in a sequence reaction to extend the known sequence (**b**). This process is continued as in (**c**), (**d**) and (**e**) until the end of the insert is reached.

(iii)   Each sequence needs the synthesis of a specific primer.

(iv)   The cost of making primers and hence obtaining the sequence is fairly high.

1.2.3 *Double-stranded sequencing*

Sequence reactions can be carried out on double-stranded templates (see Chapter 4). The quality of the results obtained is rather variable, the most important factor being the purity of the template DNA preparation. As a sequencing strategy it is most useful for checking the inserts of plasmid recombinants, since no M13 subcloning is needed. If the current quality problems can be overcome, and it is used in combination with the approach described in Section 1.2.2, using specific oligonucleotide primers, the method could prove to be very powerful for sequencing any DNA. This strategy would suffer from the same advantages and disadvantages as described in Section 1.2.2 except that none of the problems associated with M13 subcloning would apply.

1.2.4 *Ordered deletion sequencing*

There are several cloning techniques available which produce sequential deletions from one end of the insert (8–12). The deletions are normally generated enzymatically and by taking time points from the digestion the insert can be shortened in a progressive manner (as in *Figure 3*). If the deletion side of the insert is cloned adjacent to the universal primer site, then the entire insert sequence can be determined by sequencing templates from a range of time points.

Advantages:

(i)   The sequence is determined very fast.

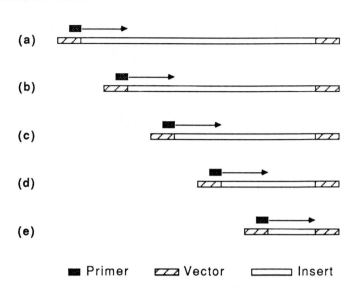

Figure 3. DNA sequencing by ordered deletions. A universal primer is used to determine the sequence, as far as possible, into the insert, as depicted by the arrow in (**a**). A series of time points of deletion (from one end of the insert) brings regions from progressively further into the insert adjacent to the universal primer site as depicted in (**b**)–(**e**).

(ii)   A specific region can be homed in on by selecting a time point.

Disadvantages:

(i)   It has to be possible to clone the entire piece of DNA in both orientations to get both strands.

(ii)   The determination of the sequence from both strands needs two separate time course experiments.

(iii)   With insert sizes in excess of 2 kb the time course becomes increasingly non-processive and the spread of start points encompassed by each time point becomes larger.

(iv)   The enzymes used to create the deletion may need prior calibration before each experiment.

(v)   There is often a requirement for flanking restriction enzyme sites and unless the insert has been characterized there is no way of knowing if any of these sites also occur internally.

1.2.5 *Random or shotgun sequencing*

Randomly produced subfragments can be produced by sonication or by digestion with DNase I (13,14). By shotgun-cloning these subfragments and sequencing the resultant recombinants at random, a complete sequence can be built up (as in *Figure 4*).

Advantages:

(i)   The sequence determination is very fast.

(ii)   No prior knowledge of the target DNA is needed.

(iii)   Any size of DNA fragment can be sequenced this way.

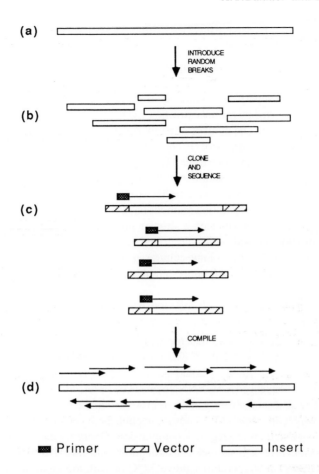

Figure 4. Random cloning and sequencing. The fragment to be sequenced in (**a**) is broken into random pieces using sonication or DNase I, (**b**). The subfragments are then cloned in M13 and sequenced at random (**c**) until the resulting data when compiled gives the entire sequence of the original fragment (**d**).

(iv)   The redundancy of data generated ensures a very high accuracy.

(v)   Size selection can be used to maximize the efficiency of gel use.

Disadvantages:

(i)   The cloning step is very inefficient.

(ii)   Data accumulation is very rapid at first but slows with the increase in redundancy.

(iii)   On average this redundancy means that eight times the original length of data is needed to complete a project totally randomly.

The simplicity of performing the DNA sequencing reactions means that many templates can be sequenced at once, a factor which makes the random strategy such a powerful method for determining large sequences very fast. The most effective means of completing any sequencing project would, however, use a combination of the strategies described above.

2. PREPARATION OF SINGLE-STRANDED TEMPLATES

The most commonly used method for purifying the recombinant M13 template DNA for sequence analysis is by growing up the individual plaques in 1.5-ml cultures and purifying the packaged viral DNA from the medium by precipitation and centrifugation from the cleared cell supernatant with polyethylene glycol (PEG) and NaCl. The viral coat proteins are removed by phenol extraction and the purified DNA ethanol-precipitated from the aqueous phase. The number of templates that can be prepared simultaneously in a single day, using this protocol, is conveniently a multiple of the number of tubes that can be placed in a microcentrifuge. This is typically 12 and therefore between 48 and 96 templates is a suitable number to prepare. Although it is possible to process 192 templates simultaneously it is a rather tedious procedure and calls for several microcentrifuges. Other preparation methods have been described, which are designed to make the automation of this step a simpler proposition (15,16) and one of these, the acid precipitation method, is also described in Section 2.4. Using these purification methods, combined with small volume cultures grown in 96-well culture plates, it should prove possible to mechanize the entire template preparation procedure.

2.1 Growth conditions for *E.coli* infected with M13

Several factors are important for the successful growth of recombinant M13 bacteriophage in *E.coli* hosts so that the template is obtained in sufficient yield and free of contaminating host DNA. *E.coli* hosts such as JM109 or TG1 (T.J.Gibson, unpublished) which are *Eco*K$^-$ should be used otherwise sequences containing the *Eco*K recognition sequence (AACNNNNNNGTGC or GCACNNNNNNGTT) will be restricted and be absent or severely under-represented in the recombinant library. Hosts such as TG2 which are *rec*A$^-$ (T.J.Gibson, unpublished) can also be used to minimize the possibility of deletions arising by recombination. Overgrowing, i.e. for longer than 5−7 h can result in cell lysis and contamination with host DNA (Chapter 3, Section 4.1.3). Growth at temperatures above 37°C or with too vigorous shaking results in lower yields of recombinant phage.

Other important factors include the following:

(i)   Plaques should be grown up and the DNA purified as soon as possible and not stored on agar plates for longer than a few days at 4°C, to prevent an increase in the background contamination on the sequencing gels.

(ii)   Care must be taken to prevent carry over of cells when transferring the phage supernatant during DNA purification as increasing amounts of host DNA may give increased non-specific priming sites.

(iii)   Do not carry over any PEG/NaCl with the pellet, which may inhibit the polymerase reaction.

(iv)   It is most important to use good quality phenol and to ensure an efficient extraction.

Good microbiological practice is essential so that contamination of cultures does not arise. It is important to have a beaker of sterilizing solution for items such as toothpicks and used tubes, and that there are traps of sterilizing solution on water pumps which are used for sucking off supernatants. It is also good practice to work on surfaces such as glass plates that are easily sterilized with ethanol after use.

2.2 Preparation of host cells and growth of recombinant M13 phage

(i) The day before, prepare an overnight culture of JM109 or TG1 cells by inoculating 10 ml of 2 × TY with a colony from a streaked minimal plate. These stock culture plates should be restreaked regularly (every two weeks).

(ii) Grow overnight at 37°C with shaking at 300 r.p.m.

(iii) The next day dilute enough of the overnight culture 1:100 with 2 × TY medium to provide sufficient for 1.5 ml per clone.

(iv) Pipette 1.5 ml of the diluted cells into sterile loose-capped 10-ml culture tubes.

(v) Without digging into the agar, toothpick a 'white plaque' (an area of infected cells which are growth inhibited) into each tube, discarding the toothpicks into a sterilizing solution.

(vi) Grow by shaking at 37°C for 5−7 h at 300 r.p.m. A New Brunswick shaker/incubator model G-24 is suitable.

2.3 Single-stranded DNA isolation by the PEG method

(i) After growing, transfer the culture to microcentrifuge tubes and centrifuge at maximum speed in a microcentrifuge for 5 min.

(ii) Carefully pour off the supernatants into new microcentrifuge tubes trying not to carry over any cells of the pellet. Make no attempt to transfer any of the remaining supernatant because of the risk of disturbing the cell pellet. Discard the original tubes containing the cell pellet into the sterilizing medium.

(iii) Add 200 μl of a solution of 20% PEG-6000, 2.5 M NaCl in water and vortex. Leave at room temperature for 10 min.

(iv) Centrifuge the tube in a microcentrifuge at maximum speed for 5 min and remove the PEG supernatant. A water pump connected to a drawn out Pasteur pipette via a liquid trap and a length of tubing is convenient for this step.

(v) Respin the tube for a few seconds to enable the removal of any remaining PEG solution. A small phage pellet should be visible.

(vi) Resuspend the pellet in 100 μl of TE buffer before adding 100 μl of TE buffer-saturated, redistilled phenol. Vortex and leave for 5−10 min at room temperature with occasional agitation. Vortex the tube again and centrifuge it in a microcentrifuge for 2 min.

(vii) Transfer most of the aqueous phase with a Gilson pipettor to a new tube taking care not to carry over any phenol. Ethanol-precipitate the DNA with 0.1 vol of 3 M sodium acetate and 2.5 vol of ethanol in a tube placed in dry ice for 20 min, at −80°C for 30 min, or at −20°C for several hours.

(viii) Centrifuge at maximum speed in a microcentrifuge for 5 min and pour off the supernatant.

(ix) Add 1 ml of 95% ethanol (there is no need to vortex), centrifuge it again for 5 min and pour off the ethanol. Leave upside down on Kleenex to drain.

(x) Dry the pellet under vacuum and dissolve it in 30 μl of TE buffer. Transfer the DNA solution to a 96-well microtitre tray for storage at −20°C.

2.4 Single-stranded DNA isolation by the acetic acid precipitation method

This method does not involve the use of either phenol extraction or ethanol precipitation and is thus more suitable for automating this process (16). It involves precipitating the bacteriophage from the cell-cleared media supernatant with acetic acid and recovering them on glass fibre filters. The phage are then dissociated on the glass fibre filters, the dissociated material washed away, and finally the DNA eluted from the filter paper with TE buffer.

(i)   After growing for 5−7 h transfer the infected cells to microcentrifuge tubes and centrifuge at maximum speed in a microcentrifuge for 5 min.

(ii)   Carefully pour off the supernatants into new microcentrifuge tubes, trying not to carry over any of the pelleted cells. Make no attempt to transfer any remaining supernatant because of the risk of disturbing the cell pellet. Discard the original tubes containing the cell pellet into the sterilizing solution.

(iii)   Add 25 μl of glacial acetic acid, cap the tube and mix. Leave at room temperature for 2 min.

(iv)   Add the suspension, 100 μl at a time, to a 7-mm diameter Whatman GF/C filter on a sintered glass filtration unit using *very* gentle suction.

(v)   Add 1 ml of 4 M $NaClO_4$ in TE buffer a drop at a time to the filter.

(vi)   Add 1 ml of 70% ethanol a drop at a time to the filter.

(vii)   Air dry the filter on parafilm for 5 min only.

(viii) Pierce a hole in the bottom of 0.5-ml microcentrifuge tubes, place a filter in each of the tubes and add 10 μl of 0.1 $\times$ TE buffer pH 7.5. Leave for 5 min at room temperature.

(ix)   Put the tube inside a 1.5-ml tube and centrifuge for 20 sec in a microcentrifuge. The eluate can be used directly in sequencing reactions.

(x)   Transfer the template solutions to microtitre trays and store as before at −20°C.

3. SEQUENCE REACTIONS

The sequencing reactions are most easily carried out in 96-well microtitre trays, dispensing the reagents in 2-μl aliquots with a Hamilton repetitive dispenser. The reagents are dispensed onto the side of the wells and mixing is achieved by briefly spinning in a bench top centrifuge fitted with a microtitre tray head such as an IEC Centra centrifuge. A more automated procedure for this, using a robot pipettor, is described in Section 4. The procedure given below uses [^{35}S]dATP as the labelled nucleotide although in practice any nucleotide could be labelled. The advantages of ^{35}S over ^{32}P are due to the lower energy of the decay particles of the ^{35}S isotope (less than one tenth the energy of ^{32}P decay) so that the bands are sharper, there is less radiolysis of the DNA and hence less background on the autoradiograph and it is intrinsically safer. The 88-day half-life compared to 14.3 days for ^{32}P also means that it has a longer shelf life and is therefore more economical if it is going to be used over an extended period. In the following protocol it is assumed that the universal sequencing primer is being used for each reaction although any oligonucleotide primer can be used. It is important to use a good quality Klenow DNA polymerase such as Boehringer Corporation sequencing grade.

3.1 Primer/template annealing

(i)   Assign wells for each reaction as shown in *Figure 5*.

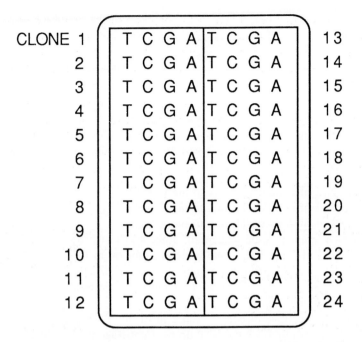

CLONE 1	T C G A T C G A	13
2	T C G A T C G A	14
3	T C G A T C G A	15
4	T C G A T C G A	16
5	T C G A T C G A	17
6	T C G A T C G A	18
7	T C G A T C G A	19
8	T C G A T C G A	20
9	T C G A T C G A	21
10	T C G A T C G A	22
11	T C G A T C G A	23
12	T C G A T C G A	24

Figure 5. Suggested layout for the sequencing reactions when they are performed manually in a 96-well microtitre plate.

(ii)   Make up enough primer/TM mix for the number of templates to be sequenced. For each template add to a 1.5-ml microcentrifuge tube 1 μl of primer (0.2 pmol), 1 μl of TM buffer and 7 μl of water.

(iii) Add 2 μl of primer/TM mix to each well using the repetitive dispenser.

(iv)   Add 2 μl of the template to the appropriate T, C, G and A wells using a Gilson P20 pipette.

(v)   Cover the microtitre plate with Saran Wrap or an adhesive plate sealer. Ensure that there are no channels through which vapour can escape, otherwise the sample will dry out.

(vi)   Centrifuge the tray briefly to mix the reagents by turning up to 2000 r.p.m. and immediately turning back to zero again.

(vii) Incubate in a 55°C oven for at least 30 min.

The annealed primer/template tray can be stored at this point at −20°C for extended periods.

3.2 Sequence reactions

(i)   If necessary, centrifuge the tray briefly to bring any condensation to the bottom of the well and then remove the Saran Wrap.

(ii)   Using the repetitive dispenser, dispense close to the rim of the well, 2 μl of the appropriate NTP mix, i.e.

   add 2 μl of T-dd/dNTP mix to the T wells,

Table 1. The volumes (in μl) of reagents in the enzyme/label mix needed for different numbers of templates[a].

	No. of templates							
	1	2	4	8	12	24	48	96
[^{35}S]dATP (10 mCi ml^{-1})	0.4	0.8	1.6	3.2	4.8	9.6	19.2	38.4
0.1 M DTT	1	2	4	8	12	24	48	96
Deionized water	7	14	28	55	80	155	305	605
Klenow polymerase (5 units μl^{-1})	0.4	0.8	1.6	3.2	4.8	9.6	19.2	38.4

[a]The volumes have been adjusted to provide a small excess.

           add 2 μl of C-dd/dNTP mix to the C wells,

           add 2 μl of G-dd/dNTP mix to the G wells,

           add 2 μl of A-dd/dNTP mix to the A wells.

(iii)   For each template to be sequenced, dilute 4 μCi of [^{35}S]dATP and 1 μl of 0.1 mM DTT, to 0.5 μCi μl^{-1} with water in a tube (as in *Table 1*) and keep on ice.

(iv)   Add Klenow DNA polymerase to the diluted [^{35}S]dATP to a final concentration of 0.125−0.25 units μl^{-1} (i.e. 1−2 units per template) as in *Table 1*, mix thoroughly and keep on ice.

(v)   Using the repetitive dispenser, add 2 μl of the ^{35}S/Klenow mix to each well taking care not to touch the NTP drops. Centrifuge the tray briefly to mix the sample and incubate at room temperature for 15 min.

(vi)   Using the repetitive dispenser, add 2 μl of the 0.5 mM dNTP chase solution to each well, centrifuge the tray to mix the reagents and incubate for a further 15 min at room temperature.

(vii)   Using the repetitive dispenser, add 2 μl of the formamide dye mix to each well and again centrifuge the tray briefly.

(viii)   Immediately prior to loading, place the tray uncovered in an 80°C oven for 15 min to denature the samples. This time should be altered if necessary so that evaporation permits application of the total sample, although prolonged heating may cause degradation (see Chapter 3, Section 4.4.4).

(ix)   Load the samples onto the gel (see Section 5) using a drawn out glass capillary or piece of tubing connected to a mouthpiece, or a narrow pipette tip and pipettor. Wash out the tip between loading samples in the bottom buffer reservoir. When using a mouthpiece, do not try to blow the last bit of the sample into the well, to minimize the risk of blowing air into the well.

3.2.1 *Replacing dGTP with dITP or 7-deaza-dGTP*

A common source of error of interpretation or unreadability on sequencing films is due to the phenomenon of compressions. These occur most frequently in GC-rich areas and are regions on the sequence ladder where the bands run anomalously and have been compressed together, quite often co-migrating. The problem is generally thought to be due to hairpin loop structures at the ends of the chains. The rationale in attempting to resolve these problems is to use nucleotide analogues in place of dGTP, which produce a base pairing stability lower than that of the G−C pair and thus lower the melting temperature of potential structures. The nucleotides are simply substituted into the nucleotide mixtures and used in otherwise normal sequencing reactions. Two such analogues, which are readily available, are inosine (dITP) and 7-deaza-dGTP (17,18). The compositions of the sequencing mixes for these analogues are given in *Table 2*. Neither of these analogues gives results of a comparable quality to those obtained when using dGTP and they are therefore used only when a compression cannot be resolved under the normal gel running conditions.

Even when these nucleotides are used, many compressed regions cannot be fully resolved in this manner. An important point to note is that compressions are artefacts which arise during the gel electrophoresis, and resolving them requires modifications to the gel environment and not to the conditions used during the sequence reactions. Another problem, commonly described as a 'pile-up', arises because of secondary structure in the template during the synthesis step. This has the appearance of heavy dark bands in all four tracks with reduced intensity, or even no ladder pattern beyond. This occurs much more rarely and can be reduced in many instances by performing the sequence reactions at elevated temperatures (even up to 50°C) in TM buffer without sodium chloride. Other techniques which can be employed to resolve compressions are using formamide in the gel composition (Section 5.2) and running the gel at elevated temperatures (Section 5.3).

3.3 Using T7 DNA polymerase or Sequenase

Klenow DNA polymerase incorporates both dNTPs and ddNTPs with differing efficiencies, as can be seen in the difference in concentrations of the dideoxy analogues needed in the nucleotide mixes. The variations in intensities of the bands on sequence films (in runs of the same nucleotide and in certain dinucleotide sequences) indicates that there are also different substrate affinities. This is particularly true in the case of

Table 2. Reagents.

---

2 × TY broth
  Make up, per litre of water:
  Bactotryptone 10 g
  Yeast extract 10 g
  NaCl 5 g
  Sterilize in 200-ml aliquots in glass bottles.

2 × TY agar
  Add 15 g l^{-1} Bacto agar to 2 × TY broth

Maintain stocks on minimal plates (see Chapter 1, Section 2).

20% PEG solution
  Polyethylene glycol 6000 20 g
  NaCl 14.6 g
  Make up to 100 ml with water.

TE buffer (pH 8.0−8.5)
  10 mM Tris−HCl (pH 8.0−8.5)
  0.1 mM Na$_2$EDTA

Table 2. (continued)

TE buffer
  10 mM Tris−HCl (pH 7.5)
  0.1 mM Na$_2$EDTA

TM buffer
  100 mM Tris−HCl (pH 8.5)
  50 mM MgCl$_2$

Phenol
  Redistill (or obtain commercially, e.g. BDH chromatographic grade) and saturate with TE buffer.

Universal 17-mer primer (GTAAAACGACGGCCAGT)
  Collaborative Research

Klenow fragment DNA polymerase I
  (5 units μl^{-1}, Cat. No. 104531): Boehringer Corporation

[α-^{35}S]dATP
  (400 Ci mmol^{-1}): Amersham or NEN

dNTPs and ddNTPs
  Pharmacia

2'-deoxy-7-deazaguanosine 5'-triphosphate
  (dc^{7}GTP or 7-deaza-dGTP): Boehringer Corporation

10 mM ddNTPs
  ddTTP, 6.1 mg ml^{-1} in TE buffer
  ddCTP, 5.8 mg ml^{-1} in TE buffer
  ddGTP, 6.2 mg ml^{-1} in TE buffer
  ddATP, 6.2 mg ml^{-1} in TE buffer

50 mM dNTP stocks
  dTTP, 31.2 mg ml^{-1} in TE buffer
  dCTP, 29.6 mg ml^{-1} in TE buffer
  dGTP, 31.6 mg ml^{-1} in TE buffer
  dATP, 29.5 mg ml^{-1} in TE buffer

0.5 mM dNTPs
  Dilute 50 mM stocks 1 in 100 in TE buffer

NTP mixes
  Mix in the following ratios:

	T	C	G	A
0.5 mM dTTP	25	500	500	500
0.5 mM dCTP	500	25	500	500
0.5 mM dGTP	500	500	25	500
10 mM ddTTP	50	−	−	−
10 mM ddCTP	−	8	−	−
10 mM ddGTP	−	−	16	−
10 mM ddATP	−	−	−	1
TE buffer	1000	1000	1000	500

DTT
  0.1 mM dithiothreitol stored at $-20°C$

dNTP chase mix
  Dilute 50 mM stocks to give
  0.5 mM dTTP
  0.5 mM dCTP
  0.5 mM dGTP
  0.5 mM dATP

Formamide dye mix
  100 ml formamide (deionized with Amberlite MB1)
  Xylene cyanol FF 0.1 g
  Bromophenol blue 0.1 g
  2 ml 0.5 M Na_2EDTA

ITP mixes
  Mix in the following ratios:

	T	C	G	A
0.5 mM dTTP	25	500	500	500
0.5 mM dCTP	500	25	500	500
2.0 mM dITP	500	500	25	500
10 mM ddTTP	50	–	–	–
10 mM ddCTP	–	8	–	–
10 mM ddGTP	–	–	2	–
10 mM ddATP	–	–	–	1
TE buffer	1000	1000	1000	500

2'-deoxy-7-deaza GTP mixes
  Mix in the following ratios:

	T	C	G	A
0.5 mM dTTP	25	500	500	500
0.5 mM dCTP	500	25	500	500
0.5 mM deaza-GTP	500	500	25	500
10 mM ddTTP	50	–	–	–
10 mM ddCTP	–	8	–	–
10 mM ddGTP	–	–	16	–
10 mM ddATP	–	–	–	1
TE buffer	1000	1000	1000	500

Nucleotide/primer/buffer mix
  (This recipe is designed for pre-filling large numbers of microtitre trays as in Section 4.)

0.5 mM dTTP	25	500	500	500
0.5 mM dCTP	500	25	500	500
0.5 mM dGTP	500	500	25	500
10 mM ddTTP	50	–	–	–
10 mM ddCTP	–	8	–	–
10 mM ddGTP	–	–	16	–
10 mM ddATP	–	–	–	1
primer (0.2 pmol μl^{-1})	250	250	250	250
T.M. (100 mM Tris–HCl, pH 8.5, 50 mM $MgCl_2$)	250	250	250	250
water	2500	2500	2500	2000

dCTP where single Cs and the first C in a run are weak. Other examples of this are a weak G in the sequence TG, a strong second C in a run of Cs, a normal intensity C when preceded by a G and a strong first A in a run of A's.

Pile ups of bands are sometimes seen opposite A bands, particularly in the sequence purine $-$C$-$A when the quality of the enzyme is poor. This is due to either the polymerase pausing, or more likely, an exonuclease stop since the thio-dATP is partially resistant to the $3'-5'$ exonuclease activity of Klenow. In this instance using a different source of enzyme, or simply adding more will often cure the problem.

Recently, chemically modified forms of T7 DNA polymerase, known commercially as Sequenase and Sequenase 2.0, have been used in sequence reactions (19). Sequenase is supplied by USB Corp. (or through Cambridge BioScience in the UK) in a kit containing the enzyme, buffers and nucleotide mixes, and is accompanied by a detailed instruction booklet for carrying out sequence reactions in microcentrifuge tubes. The unmodified enzyme is also available, from other suppliers such as Pharmacia.

T7 DNA polymerase incorporates dNTPs and ddNTPs at a more even rate and with a higher processivity than Klenow polymerase, creating a more even band pattern. Because of this a different procedure has to be used for the sequence reactions which is detailed in Section 3.3.2 and 3.3.3. As described in Section 3.1 and 3.2 for Klenow, the reactions have been adapted to make use of microtitre trays. The sequencing procedure is carried out in two steps after the primer is annealed to the template. The first is a labelling step, which is carried out in the presence of T7 DNA polymerase, [^{35}S]dATP and limiting concentrations of the other three dNTPs for around 2 min at room temperature. The primer extension is limited by low levels of the nucleotides. The relative nucleotide/DNA concentrations are important in this step, since, for example, too low a DNA concentration will result in greater extension, with consequent loss of bands at the bottom of the gel. The four separate labelling product sets then undergo the second step, chain extension and termination, carried out for 5 min at 37°C in the presence of higher concentrations of all four dNTPs and a specific ddNTP for each reaction.

The even intensity banding pattern, obtained when using T7 polymerase, is undoubtedly easier for the inexperienced to interpret. Situations where the use of T7 polymerase may be of real benefit, even to the experienced, is in the use of automatic film scanners and when fluorescent labels are used (see Chapter 8), since the signal variability is reduced and therefore automatic interpretation made significantly simpler.

3.3.1 *T7 polymerase annealing procedure*

(i)   Assign wells for each reaction as shown in *Figure 5*.

(ii) Make up enough primer/T7 buffer mix for the number of templates to be sequenced. For each template add to a 1.5-ml microcentrifuge tube 1 μl of primer (0.5 pmol), 2 μl T7 buffer and 6 μl of water.

(iii) Add 2 μl of primer/T7 buffer mix to each well using the repetitive dispenser.

(iv) Add 2 μl of the template (0.5 μg) to the appropriate T, C, G and A wells using a Gilson P20 pipette.

(v)   Cover the microtitre plate with Saran Wrap or an adhesive plate sealer. Ensure that there are no channels through which vapour can escape, otherwise the sample will dry out.

(vi)   Centrifuge the tray briefly to mix the reagents by turning up to 2000 r.p.m. and immediately turning back to zero again.

(vii) Incubate in a 55°C oven for at least 30 min.

The annealed primer/template tray can be stored at this point at −20°C for extended periods.

3.3.2 *T7 polymerase labelling procedure*

(i)   If necessary, centrifuge the tray briefly to concentrate any condensation to the bottom of the well and then remove the Saran Wrap.

(ii) For each template to be sequenced, add the following quantity of reagents to a single microcentrifuge tube:

    0.5 μl of [^{35}S]dATP (4 μCi)
    1 μl of 0.1 mM DTT
    2 μl of labelling mix
    3 units of T7 polymerase or Sequenase
    5 μl of water.

Mix thoroughly and keep on ice.

(iii) Using the repetitive dispenser, add 2 μl of the ^{35}S/T7 mix to each well taking care not to touch the droplets of template at the bottom of the well. Centrifuge the tray briefly to mix the sample and proceed immediately to the next step.

3.3.3 *T7 polymerase termination reactions*

(i)   Using the repetitive dispenser, add 2 μl of the appropriate termination mix to each well, centrifuge the tray to mix the reagents and incubate for a further 5−10 min at 37°C.

(ii) Using the repetitive dispenser, add 2 μl of the formamide dye mix to each well and again centrifuge the tray briefly.

(iii) Immediately prior to loading, place the tray uncovered in an 80°C oven for 15 min to denature the samples. The incubation time should be altered so that evaporation permits application of the total sample.

(iv) Load the samples onto the gel (see Section 5) using a drawn out glass capillary or piece of tubing connected to a mouthpiece, or a narrow pipette tip and pipettor. Wash out the tip between loading samples in the bottom buffer reservoir. When using a mouthpiece, do not try to blow the last bit of the sample into the well, to minimize the risk of blowing air into the well.

3.3.4 *Sequenase dITP reactions*

dITP mixes are provided with the Sequenase kit and the use of dITP is recommended over that of 7-deaza-dGTP. The protocol is as described in Sections 3.3.2 and 3.3.3 above, but using dITP labelling mix and dITP termination mixes (see *Table 3*). It is recommended that the dITP reactions are run alongside the normal dGTP reactions rather than replacing them because the quality of the dITP reactions is lower.

3.4 **Using Taq polymerase**

The use of Taq polymerase in amplification procedures by polymerase chain reaction

Table 3. Specialized T7 DNA polymerase solutions.

1.	T7 polymerase or Sequenase (chemically modified T7 DNA polymerase).
2.	0.5 pmol μl^{-1} Universal primer.
3.	T7 buffer (5 × concentrate)
	200 mM Tris−HCl pH 7.5
	100 mM $MgCl_2$
	250 mM NaCl
4.	Labelling mix
	2.0 μM dGTP
	2.0 μM dCTP
	2.0 μM dTTP
5.	Labelling mix for use when dITP is substituted for dGTP
	4.0 μM dITP
	2.0 μM dCTP
	2.0 μM dTTP

6.   Concentrations of the T7 termination NTP mixes

	T	C	G	A
dTTP	150 μM	150 μM	150 μM	150 μM
dCTP	150 μM	150 μM	150 μM	150 μM
dGTP	150 μM	150 μM	150 μM	150 μM
dATP	150 μM	150 μM	150 μM	150 μM
ddTTP	15 μM	−	−	−
ddCTP	−	15 μM	−	−
ddGTP	−	−	15 μM	−
ddATP	−	−	−	15 μM

7.   Concentrations of the T7 termination NTP mixes when substituting dITP for dGTP

	T	C	G	A
dTTP	150 μM	150 μM	150 μM	150 μM
dCTP	150 μM	150 μM	150 μM	150 μM
dITP	300 μM	300 μM	300 μM	300 μM
dATP	150 μM	150 μM	150 μM	150 μM
ddTTP	15 μM	−	−	−
ddCTP	−	15 μM	−	−
ddGTP	−	−	3.0 μM	−
ddATP	−	−	−	15 μM

(PCR) is now an established technique (see Chapter 4, Section 3.2). The enzyme has also been described for use in DNA sequencing (20,21), reportedly giving good results on templates which otherwise give rise to secondary structure related sequence stops. Such problems are quite rare, and the routine use of comparatively expensive Taq polymerase would be difficult to justify simply on these grounds. In addition, the use of this enzyme has two advantages over the use of Klenow polymerase. It gives (like Sequenase) a more even intensity band pattern on autoradiographs and because of its optimum reaction temperature, needs no separate annealing step.

The protocol described here, in common with those detailed above, has been designed for use in microtitre trays. The elevated temperature of the Taq sequencing reactions and their short incubation times make a microtitre tray format heat block very useful. A cheaper alternative is to make a small aluminium block with 96 shallow drill holes. This can be placed on top of a standard heat block.

3.4.1 *Taq polymerase sequencing protocol*

(i)   Assign wells for each reaction as shown in *Figure 5*.

(ii)   Add 2 μl of the template (0.5 μg) to the appropriate T, C, G and A wells using a Gilson P20 pipette.

(iii)   Make up enough primer/Taq/buffer (*Table 4*) mix for the number of templates to be sequenced. For each template add to a 1.5-ml microcentrifuge tube:

   1 μl of primer (0.5 pmol)
   2 μl of Taq buffer
   2 μl of labelling mix
   0.5 μl of ^{35}S dATP (400 Ci mmol^{-1})
   11 μl of water
   1 unit of Taq polymerase.

(iv)   Add 4 μl of primer/Taq/buffer mix to each well using the repetitive dispenser. It should not be necessary to centrifuge the tray as the detergent in the Taq buffer will prevent the droplet adhering to the side of the well.

(v)   Place the tray on a heat block at 70°C for 2 min. There is no need to cover the microtitre plate unless the heat block is positioned in air currents.

(vi)   Using the repetitive dispenser, add 2 μl of the appropriate termination mix to each well and incubate for a further 5 min at 70°C on the heat block.

(vii)   Using the repetitive dispenser, add 2 μl of the formamide dye mix to each well.

(viii) Immediately prior to loading, place the tray, uncovered, on the heat block at 95°C for 5 min to denature the samples. The incubation time should be altered so that evaporation permits application of the total sample.

(ix)   Load the samples onto the gel (see Section 5) using a drawn out glass capillary or piece of tubing connected to a mouthpiece, or a narrow pipette tip and pipettor.

Table 4. Specialized Taq polymerase solutions.

1.	Taq polymerase.			
2.	0.5 pmol μl^{-1} Universal primer			
3.	Taq buffer			
	50 mM Tris−HCl pH 7.5			
	50 mM MgCl$_2$			
	0.5% Tween 20			
	0.5% NP40			
4.	Labelling mix			
	10.0 μM dGTP			
	10.0 μM dCTP			
	10.0 μM dTTP			
5.	Taq termination mixes:			
	T	C	G	A
dTTP	30 μM	30 μM	30 μM	30 μM
dCTP	30 μM	30 μM	30 μM	30 μM
dGTP	30 μM	30 μM	30 μM	30 μM
dATP	30 μM	30 μM	30 μM	30 μM
ddTTP	1.5 mM	−	−	−
ddCTP	−	0.5 mM	−	−
ddGTP	−	−	0.25 mM	−
ddATP	−	−	−	1.0 mM

Wash out the tip between loading samples in the bottom buffer reservoir. When using a mouthpiece, do not try to blow the last bit of the sample into the well, to minimize the risk of blowing air into the well.

4. AUTOMATION OF DNA SEQUENCING

4.1 **The sequencing machine**

Several reports have been published proposing or describing equipment designed to automate the entire process of DNA sequencing (22,23). One ambitious proposal is that of a Japanese consortium who are planning to mechanize the current manual radioactive process using robots and mass produced pre-cast gels. For those researchers who view automation simply as a means of obtaining DNA sequences more quickly, more easily and at reasonable cost, there are a few major drawbacks to such a machine which is designed to carry out the whole process automatically:

(i)   A fully automated instrument is clearly going to be very expensive and only a small proportion of molecular biologists who are currently interested in using the technique of DNA sequencing will want to do it on a large scale, let alone be able to justify and raise the capital needed to purchase it.

(ii)   For many, the only possibility of employing such a fully automated machine would be if it were purchased by departments and installed as a service unit. To be useful under these circumstances, its throughput would need to be at least as high as can currently be achieved by the number of individuals it will be serving. At present manual rates, this can be as high as $40-60$ templates per day for a single researcher. The currently proposed devices, in comparison, have a very low capacity.

(iii) It is not possible to predict when technology changes or new procedures will leave an expensive and totally outdated piece of hardware. It is obvious from the scale of some of the projects being proposed that current technologies and procedures are totally inadequate and to achieve their aims major advances are not only desirable but essential.

In view of these requirements, there are distinct advantages to be gained in retaining a modular approach to carrying out DNA sequencing and settling for a partial automation of each module. Not only could the separate modules be amended as required, but each could be designed at low cost, or designed to be sufficiently flexible to be used in other processes. The capability of employing current commercially available equipment would also be of great advantage.

When viewed in a modular fashion, DNA sequencing can be split into four procedures:

(i)   Cloning/template preparation.
(ii)   Performing sequence reactions on the templates.
(iii) Gel electrophoresis of the samples.
(iv)   Interpretation and compilation of the data.

By treating these modules independently, it is possible to automate highly most aspects of the technique and indeed several automated approaches to the modules described above are already being developed.

One very feasible set-up would use a highly purified double-stranded template (so that a single preparation is required) which could be subjected to sequence analysis using a synthetic oligonucleotide primer, by a robot as described in Section 4.2. If the reaction products are fluorescently labelled, they can be simultaneously fractionated and analysed to the limit of the resolution of the detector. In turn, this sequence output would direct the automatic synthesis of a new primer oligonucleotide, which would hybridize close to the limit of the known sequence. This stepwise progress could continue to the end of the entire fragment. Since only a single step at a time is possible, to maximize the rate of sequencing several such fragments would be processed concurrently, or several start points would be used.

4.2 **Automation of sequence reactions**

Of the various steps involved, the performing of the DNA sequence reactions is probably the simplest routine to automate. The multi-pipetting of small volumes of reagents to specific reaction vessels at measured time intervals is all that is required. Using the current manual methods described in Section 2, all the additions have been standardized to a volume of 2 μl. This volume is small enough to minimize the total volume of the reagents, but large enough to make the additions quick and simple. Smaller volumes are often difficult to dispense from diposable polypropylene tips, especially when viscous solutions are involved. The standardization of reagent volume also permits the use of cheap repetitive dispensers such as the Hamilton PB 600. Using these, fitted with a 100-μl syringe and disposable tip, each depression of the repeating mechanism dispenses 2 μl of reagent which can be released into the well by touching the droplet onto the side. The individual droplets are driven to the well bottom and mixed by rapid acceleration to 1500 r.p.m. and immediate deceleration in a bench top centrifuge.

This technique has been used successfully in this laboratory to sequence the two largest contiguous stretches of DNA sequences yet determined (the herpes viruses Epstein – Barr virus, 170 kb, and Human Cytomegalovirus, 230 kb). This practical experience, as opposed to theoretical calculation, has shown that using manual methods a single researcher can realistically aim to determine around 50 kb of contiguous, completely double-stranded sequence in a year. In our hands this has, as yet, not been extrapolated to 100 kb in 2 years, highlighting a more practical reason for pursuing automation, the tedious nature of the task. Microtitre plates were introduced as the reaction vessel, not only because they facilitated the handling of many sequence reactions simultaneously, but also the potential for using already available microtitre based dispensers/diluters for automation was recognized.

There are several robotics-based pipetting instruments currently available which are capable of doing just this. Designed initially for automating ELISA, many do not cope well with the very small volume dispensing capabilities needed for DNA sequencing and the single tip dispensing of others makes large numbers of additions very slow. All of the machines tested by us which are capable of delivering volumes less than 3 μl have, however, been able to perform sequence reactions. The Beckman Biomek 1000 has been found to be capable of dispensing volumes down to 1 μl into microtitre plates using an eight tip dispensing tool which uses disposable tips. Its self changing, multi-tool capability also makes it extremely flexible in the range of possible applications for which it may be useful and it therefore need not be dedicated to DNA sequencing.

Using the Beckman Biomek 1000 for DNA sequencing involves only minor changes to the currently used manual protocol described in Section 3. It can, of course, be used to perform reactions using any of the available polymerases or fluorescently-labelled nucleotides. A revised reaction plate layout and reagent storage system is all that is needed. One of the three plate positions on the tablet is used as reagent reservoir and the other two are available for reaction plates. Complete sequence reactions, from combining different templates with primer, through to adding formamide, can be executed on the Biomek using simple pipetting routines.

The rate of completing a pipetting task, using robots, is largely limited by the speed at which the motors can be moved. Since, when sequencing, reagent additions need to be at specific time intervals, only a limited number of individual steps are possible before the next addition has to be started. The use of an eight tip dispensing tool greatly speeds this up, but in attempting a considerable scale-up of the rates at which sequencing is carried out, a revised approach is needed to the order in which the various parts are performed. In the same way that the whole process of DNA sequencing can be considered as separate procedures, the sequence reactions themselves can be split into component parts. The individual parts of the process can be done separately, in bulk, removing any timing constraints, avoiding the need for linking together complex subroutines and maximizing the efficiencies of the component steps, as robots are especially well suited to performing simple repetitive tasks. Convenient divisions for this splitting into components are:

(i)   Plate fill with primer, buffer and nucleotides.
(ii)   Template DNA addition and annealing.
(iii)   Sequence reactions.

Programming the Biomek is a fairly simple task. The program is built up from functions (the basic sample handling routines) which are linked together within a subroutine. These subroutines can be linked together making functions which in turn can be combined to give the complete method. Up to four different tools can be stored on the tablet and can be exchanged automatically as the working tool, within a method. For the methods described here, only the eight tip MP20 tool is used.

Using a microtitre plate as the reagent vessel (bulk volume quarter trays can be used if many trays are being filled) 4 μl of combined primer/buffer/nucleotide solutions (reagent compositions are detailed in *Tables 2* and *3*) are dispensed as in *Figure 6* and outlined in *Table 5*. These stock reagent plates can be stored for at least several months at $-20°C$ until needed. The potential for commercially produced plates is obvious. The samples can even be dried down in the wells to ease transportation. Several of these dried down reagent plates have been transported long distances at ambient temperature, subsequently stored at $-20°C$, and used in robotic sequencing with little deterioration.

Template DNAs for sequence reactions are currently stored in microtitre plates sealed with standard acetate plate sealers. They are economical of freezer space and ideal for storing large numbers of templates. If a plate well/template numbering format of plate position A1 = 1 through to H12 = 96 is used, setting up the annealing of 48 templates is performed by using the routine detailed in *Table 6* and shown schematically in *Figure 7*. Copying currently used protocols, these plates are covered with Saran Wrap or acetate plate sealers to minimize evaporation, and placed at 55°C for 30 min to anneal. The annealing step can also be carried out at ambient temperature for about 1 h *in situ*.

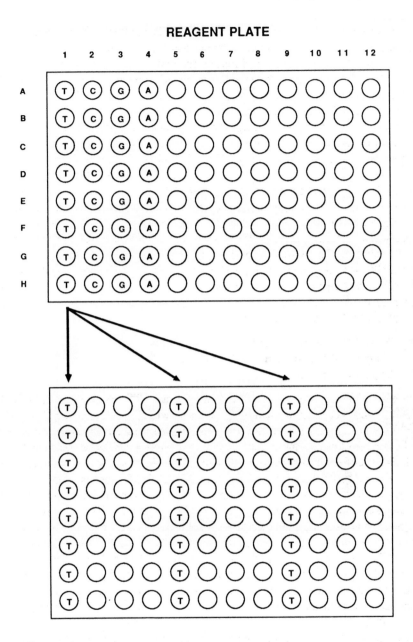

Figure 6. Using a Biomek 1000 to prepare stock reaction plates for DNA sequencing. Four pre-mixed nucleotide/primer/buffer solutions are stored in a reagent plate or strip wells. The Biomek 1000 is then used to dispense the solutions into the correct wells of many reaction plates. As many plates as required can be prepared in this manner and stored at −20°C for use even months later.

Table 5. Biomek method used to fill trays 2 and 3 with nucleotides/primer/buffer solutions.

Configuration

Tips	P250	
Tray 1	96-well	Vinyl-U (reagent)
Tray 2	96-well	Vinyl-U
Tray 3	96-well	Vinyl-U
Tool A	MP 20	

Subroutine A (T solution)

Function		1	2	3
	Volume	4	4	4
	Rate	4	4	4
	Source	Tray 1 A1-H1	Tray 1 A1-H1	Tray 1 A1-H1
	Dest.	Tray 2 A1-H1	Tray 2 A5-H5	Tray 2 A9-H9
	Source hgt.	Bottom	Bottom	Bottom
	Dest. hgt.	Bottom	Bottom	Bottom
	Contain	√	√	√
	Blowout	√	√	√
	Tip touch at dest.	√	√	√
	Tip change	−	−	−
	Pre-wet	−	−	−

Subroutine B (T solution)

Function		1	2	3
	Volume	4	4	4
	Rate	4	4	4
	Source	Tray 1 A1-H1	Tray 1 A1-H1	Tray 1 A1-H1
	Dest.	Tray 3 A1-H1	Tray 3 A5-H5	Tray 3 A9-H9
	Source hgt.	Bottom	Bottom	Bottom
	Dest. hgt.	Bottom	Bottom	Bottom
	Contain	√	√	√
	Blowout	√	√	√
	Tip touch at dest.	√	√	√
	Tip change	−	−	−
	Pre-wet	−	−	−

Subroutine C Tip change
The function set-up continues as in subroutines A and B except:

Subroutine D (C solution)

Function		1	2	3
	Source	Tray 1 A2-H2	Tray 1 A2-H2	Tray 1 A2-H2
	Dest.	Tray 2 A2-H2	Tray 2 A6-H6	Tray 2 A10-H10

Subroutine E (C solution)

Function		1	2	3
	Source	Tray 1 A2-H2	Tray 1 A2-H2	Tray 1 A2-H2
	Dest.	Tray 3 A2-H2	Tray 3 A6-H6	Tray 3 A10-H10

Subroutine F Tip change

Subroutine G (G solution)

Function		1	2	3
	Source	Tray 1 A3-H3	Tray 1 A3-H3	Tray 1 A3-H3
	Dest.	Tray 2 A3-H3	Tray 2 A7-H7	Tray 2 A11-H11

Subroutine H (G solution)

Function		1	2	3
	Source	Tray 1 A3-H3	Tray 1 A3-H3	Tray 1 A3-H3
	Dest.	Tray 3 A3-H3	Tray 3 A7-H7	Tray 3 A11-H11

Subroutine I Tip change
Subroutine J (A solution)

Function		1	2	3
	Source	Tray 1 A4-H4	Tray 1 A4-H4	Tray 1 A4-H4
	Dest.	Tray 2 A4-H4	Tray 2 A8-H8	Tray 2 A12-H12

Subroutine K (A solution)

Function		1	2	3
	Source	Tray 1 A4-H4	Tray 1 A4-H4	Tray 1 A4-H4
	Dest.	Tray 3 A4-H4	Tray 3 A8-H8	Tray 3 A12-H12

Table 6. Biomek method used to dispense template DNA to trays 2 and 3.

Configuration

Tips	P250	
Tray 1	96-well	Vinyl-U (templates)
Tray 2	96-well	Vinyl-U
Tray 3	96-well	Vinyl-U
Tool A	MP20	

Subroutine A (templates 1−24)

Function		1	2	3
	Volume	2	2	2
	Rate	4	4	4
	Source	Tray 1 A1-H1	Tray 1 A2-H2	Tray 1 A3-H3
	Dest.	Tray 2 A1-H4	Tray 2 A5-H8	Tray 2 A9-H12
	Source hgt.	Bottom	Bottom	Bottom
	Dest. hgt.	25%	25%	25%
	Contain	√	√	√
	Blowout	√	√	√
	Tip touch at dest.	√	√	√
	Tip change	−	−	−
	Pre-wet	−	−	−

Subroutine B Tip change

Subroutine C (templates 25−48)

Function		1	2	3
	Volume	2	2	2
	Rate	4	4	4
	Source	Tray 1 A4-H4	Tray 1 A5-H5	Tray 1 A6-H6
	Dest.	Tray 3 A1-H4	Tray 3 A5-H8	Tray 3 A9-H12
	Source hgt.	Bottom	Bottom	Bottom
	Dest. hgt.	25%	25%	25%
	Contain	√	√	√
	Blowout	√	√	√
	Tip touch at dest.	√	√	√
	Tip change	−	−	−
	Pre-wet	−	−	−

TEMPLATE DNA

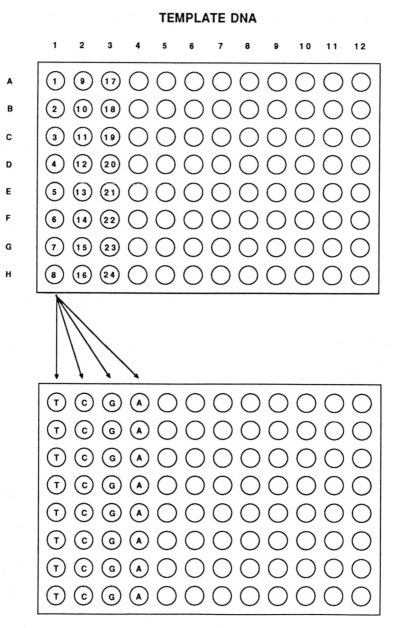

REACTION PLATE

Figure 7. Using the Biomek 1000 to dispense template DNA. Single-stranded template DNA stored in microtitre trays is dispensed into the appropriate wells of a pre-prepared reaction plate already containing the nucleotide/primer/buffer mixture.

After cooling to room temperature the plates can be used immediately in sequence reactions or, if they are prepared in bulk for future use, can be stored at −20°C for at least several months.

Three reagents are needed to complete the sequence reactions on the annealed templates.

(i) ^{35}S/DTT/Klenow: 50 μCi [^{35}S]dATP, 12 μl of 0.1 M DTT, 25 units of Klenow pol I per 100 μl.

(ii)   Chase solution: 0.5 mM in all four dNTPs.

(iii) Formamide dye mix: 99 ml of deionized formamide, 1 ml of 0.5 M EDTA, 0.1 g of xylene cyanol, 0.1 g bromophenol blue.

These three reagents are positioned in the first three columns of the reagent microtitre plate. The method of DNA sequencing is then simply a plate fill routine with incubation time gaps between each addition, using the routine in *Table 7*. Again at this stage, samples can be stored for running on gels at a later date, but the presence of formamide can sometimes result in sample degradation. If storage of the plates is planned, then it is best to omit adding the formamide and to add it immediately before denaturation and loading of the samples onto gels.

Although it has not proven to be a problem in our hands, one difficulty foreseen with this layout is the possible instability of the enzyme at ambient temperatures. This problem can be minimized by using strip wells and storing the enzyme reagent strip at −20°C, positioning it in the reagent tray only immediately before use. In fact, the reagents are more convenient when stored in strip wells, as they can be quickly replaced when they run out. Again the potential for commercially produced consumables can be seen.

5. GEL ELECTROPHORESIS

5.1 Polyacrylamide gels

Acrylamide is an aminocarbonyl compound which, in aqueous solution in the presence of the initiators ammonium persulphate and N,N,N',N'-tetramethylethylenediamine (TEMED), can undergo a polymerization reaction. The free radical mediated polymerization produces long chains of polyacrylamide and continues until the monomer is exhausted. Intertwining of these polymers produces a viscous gel of low mechanical strength. If N,N'-methylene bis-acrylamide is included in the monomer mixture at comparatively low concentrations ($<5\%$), upon polymerization it forms branches which can produce covalent bridges between the polyacrylamide chains. This not only imparts strength to the gel, but gives a meshwork with an effective mean pore size which can be varied by altering the total concentration of monomers.

Mixtures of macromolecules, including DNA, can be fractionated through polyacrylamide by electrophoresis. The mixture is placed into slots moulded in the gel at one end and an electric field is applied across the length of the gel. Molecules which carry charge migrate under the influence of such a field at a rate dependent upon their total charge. DNA, which has negative charge, moves towards the anode. In the polyacrylamide, the migration is retarded by the polymer meshwork in a fashion which is more dependent upon molecular size. In the case of DNA chains, this is length. The shorter molecules experience less resistance to movement and migrate faster.

Table 7. Biomek method used to perform sequence reactions on annealed templates/primer in tray 3.

Configuration		
Tips	P250	
Tray 1	96-well	Vinyl-U (reagent)
Tray 3	96-well	Vinyl-U (templates/primer)
Tool A	MP20	

Subroutine A (enzyme/radiolabel)		
Function		1
	Volume	2
	Rate	4
	Source	Tray 1 A1-H1
	Dest.	Tray 3 A1-H12
	Source hgt.	Bottom
	Dest. hgt.	25%
	Contain	√
	Blowout	√
	Tip touch at dest.	√
	Tip change	√
	Pre-wet	—

Subroutine B   Pause 20 minutes

Subroutine C (nucleotide chase)		
Function		1
	Volume	2
	Rate	2
	Source	Tray 1 A2-H2
	Dest.	Tray 3 A1-H12
	Source hgt.	Bottom
	Dest. hgt.	25%
	Contain	√
	Blowout	√
	Tip touch at dest.	√
	Tip change	√
	Pre-wet	—

Subroutine D   Pause 15 minutes

Subroutine E (formamide dye solution)		
Function		1
	Volume	2
	Rate	4
	Source	Tray 1 A3-H3
	Dest.	Tray 3 A1-H12
	Source hgt.	Bottom
	Dest. hgt.	25%
	Contain	√
	Blowout	√
	Tip touch at dest.	√
	Tip change	√
	Pre-wet	—

The products of radioactively-labelled sequence reactions are four, nucleotide-specific, mixtures of partially double-stranded DNA molecules. These molecules consist of one large template strand and one radioactively-labelled, primer-extended strand, which is of variable length. The two strands of the sequence reaction products are denatured immediately before application onto the polyacrylamide gel by the addition of formamide and heating the samples to between 80°C and 100°C. The heating reduces water content by evaporation, which serves two purposes. Firstly the formamide concentration increases, aiding denaturation, and secondly the total sample volume is reduced. The smaller sample volume means that more DNA can be applied onto the small loading slots of sequencing gels. When the electric field is applied, the single-stranded DNA molecules migrate towards the anode. The much larger template strand moves only very slowly, and remains close to the sample application wells. The primer-extended molecules, having different lengths, travel at different rates through the gel, the shorter ones moving faster, and so are separated. They can then be visualized by autoradiography.

5.2 Sequencing gels

As well as intermolecular base pairing between the template and the primer extension products, intramolecular base pairing can arise, forming small loop structures within the extension products. This introduces conformational mobility differences to the relatively simple principle assumed above of fractionation according to length. On sequencing gels the manifestation of these is the compression, where the mobility of a fragment is altered to be the same as one a few nucleotides shorter as a result of a short region of potential base pairing at its 3′ end. The name was coined because of its appearance on autoradiographs, where the spacing of the bands is compressed locally as a result of the mobility change. Including urea in the gel mix or replacing some, or all, of the water with deionized formamide, and maintaining a high gel temperature during the run both help to avoid these problems by destabilizing the secondary structure. Heat energy produced by electrophoresis is sufficient by itself to keep a high running temperature. A recipe for 6% acrylamide/urea gel mix is detailed in *Table 8*.

It is quite important that there is a uniform electric field across the width of the gel as otherwise the resulting band pattern will be distorted and lack resolution. Using two sheets of float glass and a uniform thickness spacer material to make the mould ensures an even thickness of gel. To enable electrode buffer contact with the gel, one of the glass plates has a notch cut out. Although these plates are comparatively expensive to produce, they are convenient and, if handled carefully, will last for several years. A cheaper alternative is to use one shorter plate to which small glass 'ears' are cemented. Before assembly the notched plate is treated with a silanizing solution which encourages the gel to stick to the other glass surface when it is being taken apart at the end of the run. In practice, the gel seems to adhere to either with approaching equal frequency, but the smoother surface does make gel pouring easier and for this reason silanizing is recommended.

To maximize the resolution of the gel system very thin gels are used. A cheap and convenient spacer material is Plastikard (polystyrene) which is sold as modeller's card. Various thicknesses are available, but the 0.015 inch (0.35 mm) thick sheet gives a

Table 8. Acrylamide gel electrophoresis reagents.

---

10 × TBE
  Tris 108 g
  Boric acid 55 g
  Na$_2$EDTA 9.3 g
  Dissolve and make up to 1 litre with deionized water

40% acrylamide solution
  Acrylamide (Electran BDH) 380 g
  N,N'-methylene bisacrylamide (Electran BDH) 20 g
  Make up to 1 litre with deionized water
  Stir gently with 20 g of mixed bed resin (e.g. Amberlite MB1) for 5−10 min
  Filter through sintered glass to remove resin
  Store at 4°C

0.5 × TBE 6% gel mix
  75 ml 40% acrylamide
  25 ml 10 × TBE
  Urea (BRL ultrapure) 230 g
  Make up to 500 ml with deionized water
  Filter through sintered glass
  Store at 4°C for up to a month

5.0 × TBE 6% gel mix
  30 ml 40% acrylamide
  100 ml 10 × TBE
  Urea (BRL ultrapure) 92 g
  (Bromophenol blue 10 mg) optional
  Bring volume to 200 ml with deionized water
  Filter through sintered glass
  Store at 4°C for up to 1 month

25% AMPS
  25% ammonium persulphate in water
  Can be stored for several months at 4°C

TEMED
  N,N,N',N'-tetramethyl-1,2-diaminoethane

Dimethyldichlorosilane solution
  BDH (for silanizing gel plates)

---

good compromise between band resolution and ease of sample loading. Two strips of this material are used, one down each edge, to space the two sheets of glass. The same material is used to form the sample wells at the top of the gel. A comb shaped piece of Plastikard is cut or, preferably, machined to give teeth of the desired well dimensions. If only a few slot formers are being made, the teeth can be cut by pressing down onto a safety razor blade positioned on the sheet of card. When more than just a few are needed, or if very small wells are used, many slot formers can be fashioned simultaneously by using a milling machine to make cuts vertically into a stack of blanks held edge upward. This method of cutting gives a flat surface to the holes cut in the comb and therefore a square top to the well dividers of the polymerized gel. Immediately after pouring the gel mix into the mould, the slot former is inserted so that once the gel has polymerized, and the slot former is removed, convenient loading wells are cast in the gel.

The size of the wells is largely a matter of personal preference; smaller wells mean more samples per gel at the expense of ease and speed of sample loading. Since doing sequence reactions in microtitre plates forces the use of an 8×12 (24 clones) format, it is logical to take some fraction of this as the number of samples loaded onto each gel. As it turns out, 48 samples (12 clones) fit comfortably onto a 20 cm wide gel without causing too many difficulties for a practised gel loader. The slot former for this format has around 50 'teeth' 2 mm wide and 3 mm deep with gaps of 1 mm in between. The actual loading width of the gel is limited to around 16 cm; this slot size avoids using the very edge of the gel and still allows the casting of a few extra slots in case some do not form properly. Overall the slot former has a width of 15 cm and a depth of $40-50$ mm so that it can be gripped for removal.

5.2.1 *Preparation of a sequencing gel*

(i)   Thoroughly clean the glass plates in warm water and rinse them with distilled or deionized water. If the plates are cleaned in this way immediately after use, and no gel is allowed to dry onto the surface, there is no need to use abrasives or detergents.

(ii)   After drying the plates, and working within a fume cupboard, silanize the smaller or eared plate by spreading 2 ml of silanizing solution over all of one surface (the one which will be the inner surface) with a Kimwipe tissue and leave it to air dry.

(iii)   Wipe both plates with a few ml of 95% ethanol using Kimwipes, polishing them well. This greatly aids the pouring of the gel into very narrow moulds.

(iv)   Assemble the gel mould using the two glass plates and two side spacers and form a liquid-tight seal around its edge by wrapping with polyester tape across the gap.

(v)   Allow sufficient gel mix to warm to room temperature. The actual volume needed will depend upon the dimensions of the glass and the precise thickness of the spacer material (a gel which is 20 cm wide $\times$ 40 cm long $\times$ 0.35 mm thick needs around 40 ml, whereas a 50 cm long gel of the same other dimensions needs around 45 ml) but it is always better to allow an excess for the possibility of leakage and spillage.

(vi)   For each ml of gel solution, add 2 μl of 25% (w/v) ammonium persulphate and then 2 μl of TEMED, with stirring, and immediately pour the mix into the gel mould. This is most conveniently done using a disposable 50-ml syringe, which gives good single-handed control of the rate of adding the gel mix. The rate of flow of gel mix within the glass mould is controlled by altering the angle at which the plates are held in the other hand.

(vii) When the mould is full, insert the slot former to a depth of approximately 0.5 cm and clamp the edges of the plates, top and bottom, directly over the side spacers, to ensure an even thickness of gel.

(viii) Leave the gel nearly horizontal to polymerize for 30 min, topping up with the remaining gel mix if the level drops.

Another popular type of slot former, called a 'sharkstooth' (which gets its name from the pointed tooth shape), is positioned onto a flat, polymerized gel loading surface with the teeth just penetrating the gel. The former is left in place during loading and is itself used as the barrier between samples. Each sample is applied between the teeth of the former. The attraction of these is that the sample tracks are immediately adjacent because of the tapering of the teeth. This makes interpretation of the resulting autoradiograph banding pattern simpler because the exact order of the bands is more clear. The disadvantages are that a perfectly flat gel surface is needed, the barrier between samples often leaks and application of samples is made more difficult by the body of the former. All of these disadvantages become more severe when extremely small slots are used.

5.3 Gel running apparatus

Uneven dissipation of heat across the width of the glass plates can distort the electric field due to resistance changes. This is seen in the autoradiograph as 'smiling' of the band pattern. The samples close to the side spacers curve upwards as migration rates are slowed in an area of lower voltage drop. This type of overall distortion generally poses little or no problem to the experienced gel reader, but can be a severe limitation to some automatic scanning systems. The problem can be resolved in several ways.

(i)   Using narrow side spacers (<0.5 cm) and avoiding the gel area within around 2 cm of the edge of the gel; this is the part most affected by the temperature gradient.

(ii) Placing a good conductor of heat, such as aluminium, in direct contact with the outer glass plate; this distributes the heat loss more evenly.

(iii) Using a purpose built gel running apparatus which has either passive or active temperature control.

Those apparatuses employing passive temperature control generally use one of the electrode buffer solutions as a heat sink, having it in contact with a large proportion of one of the gel mould surfaces. Using active temperature control, on the other hand, a heating/cooling solution, which comes from a temperature-controlled reservoir, is pumped around a chamber in contact with one of the gel mould surfaces or even around a specifically made gel mould surface in direct contact with the gel.

Active temperature control does allow the gels to be run marginally faster, the excess heat generated is removed by the cooling solution, and they do permit precise control of elevated temperatures (even up to 90°C), by increasing the temperature of the reservoir so that it heats the gel. This is useful when particularly strong intramolecular structures need to be resolved. These purpose built thermostatting gel tanks are very expensive and can be as much as ten times the cost of a simple functional design which is perfectly adequate under nearly all circumstances. A simple two tank design, as shown in *Figure 8*, also gives added height flexibility, since the single apparatus can be used for any length of gel, the gel mould itself supporting the upper tank.

5.3.1 *Preparing the gel for loading*

(i)   Wash away any polymerized gel from the outside of the gel plates and slowly remove the slot former under an overlay of deionized water.

(ii) Remove the sealing tape from the bottom of the plates or slit it with a razor blade.

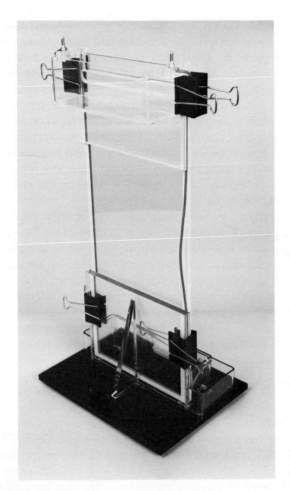

Figure 8. A simple universal length vertical electrophoresis system. A basic design 20 cm high electrophoresis apparatus has been sawn in two and the lower electrode cable extended. The two halves of the system are clamped to a standard glass plate gel mould. The mould supports the upper tank and can be of any length. Gel electrophoresis should be carried out with an appropriately sized safety cover on (not shown) or in a purpose built safety cabinet.

(iii)   Position the gel plates into the bottom chamber a little above the base with the notch facing the back and hold it to the body of the chamber with foldback clips.

(iv)   Using foldback clips clamp the upper buffer chamber to the plates so that the lip of the chamber is level with the lower edge of the notch.

(v)   Fill both the chambers with TBE and immediately flush the wells of the gel, using a Pasteur pipette or a syringe and needle, to wash away any unpolymerized acrylamide and urea.

5.4 Gel running

Immediately after denaturation a volume of less than 2 μl of each reaction has to be loaded into the sample wells. This takes a little skill and practice, as the whole gel needs to be loaded within about 10 min to minimize renaturation and diffusion of samples. The simplest but slowest method is to use a standard small volume pipette fitted with a disposable polypropylene tip. These afford poor control of the sample flow and because the tip is much wider than the thickness of the gel the sample cannot be pipetted directly into the well bottom. The greater density of formamide is relied upon to make it trickle to the bottom which causes inevitable dilution with buffer and trailing. Polypropylene tips which are flattened or have a small outside diameter are commercially available, which help to some degree. One supplier, Drummond, markets a pipette designed for loading sequencing gels. It has a low volume capacity and is fitted with a replaceable fine point polycarbonate capillary tip. No matter which pipette is used, between each loading the tip has to be changed or flushed to avoid sample cross-contamination, considerably slowing down the loading process.

The technique preferred by many sequencers, but actively discouraged by safety groups, is mouth pipetting. A capillary similar to those of the Drummond pipette or made by 'pulling' narrow bore glass or polypropylene tubing in a cool flame to make a fine point is connected via a length of tubing to a mouthpiece. These give very precise control of sample flow and the capillary can be quickly flushed between samples simply by blowing briefly whilst it is submerged in the anode buffer (not at the cathode, which would cause a background in the autoradiograph).

With four different reactions constituting a set of sequence products, there are only a limited number of permutations for the order of loading of samples. Everyone seems to have their own preference, and almost all the possibilities are used by one person or another. There are, in fact, two good reasons for choosing specific track orders. Firstly, to facilitate the reading of films in either direction (since inserts can be ligated into the vector in either of the orientations) it is convenient to have a track order where simply flipping over the film and reading it from top to bottom will give the complementary sequence. This is the case if the pairs of complementary nucleotide tracks occupy the two centre lanes for one pair and the two outside lanes for the other (A + T in either of the two centre lanes, or G + C in either of the centre lanes). Secondly, the most common sequence interpretation problems are associated with compressions. These nearly always arise because of G−C pairing and the interpretation difficulties arise because there is doubt about the precise order of these G and C bands. It makes sense, therefore, to have these tracks immediately adjacent. Combining these considerations, there are four preferred orders, TCGA, TGCA, ACGT, AGCT. None of these has an advantage over the others, TCGA has historical origins as a main choice but many use the alphabetical ACGT.

5.4.1 *Loading and running the gel*

(i)   Heat denature the sequence reaction products in the microtitre tray in an oven at 80°C for 20 min, or until the sample volume has been reduced to around 3 μl.

(ii) Flush the loading wells with TBE, using a Pasteur pipette or a syringe and needle, to remove any urea which has diffused out into the wells.

(iii) Take up into the loading pipette or capillary 2 μl of sample, avoiding introducing air.

(iv) Load the sample into the well close to the bottom maintaining a smooth flow.

When using a mouthpiece apply only 90% of the sample in the capillary to avoid blowing air through and spreading the sample.

(v)   Immerse the capillary or tip into the anode buffer or a beaker of water and rinse it to remove most of the remaining sample.

(vi)   Load all of the wells in this way as quickly as possible to prevent possible renaturation and diffusion of sample.

(vii)   When the gel is fully loaded, put on the cover or place it in a safety cabinet and connect it to the power supply with the positive terminal at the bottom of the gel. Run the gel for the required time with the supply set on constant power (this ensures a constant heat input).

Using the 6% acrylamide gels described in Section 5.7 (linear, wedge or gradient) the bromophenol blue marker runs at an approximately equivalent rate to a 25 nucleotide long single-stranded length of DNA (including the primer) and the xylene cyanol marker at an equivalent rate to approximately 115 nucleotides. The actual run time required will depend upon the length of the primer molecule and the distance from its position of hybridization to the point where the start of sequence data is desired. For 0.35 mm thick 20 × 50 cm gels use a power of 35−40 W and run them for around 2.5 h to get the bromophenol blue marker dye to the bottom of the gel (40 cm long gels need around 30−35 W and will take around 2 h to run the bromophenol blue to the bottom).

5.5 Autoradiography

The radioactively-labelled DNA bands within the gel are detected by autoradiography. A sheet of X-ray film (Fuji RX or Kodak XAR) is placed in close contact with the gel. After an exposure period of between 12−48 h the film is processed and a clear banding pattern should be seen. The sharpness of the autoradiograph bands, as well as being determined by the band sharpness within the gel, are also affected by the autoradiographic technique. If the film is not in intimate contact with the radioactive source then the bands will be diffuse, since the radioactive particles radiate from the source. When such sources as ^{32}P with powerful emissions are used, the radiant energy is detected at comparatively large distances and radioactive decay from deeper in the gel gives rise to a wider spread of detection and therefore a more diffuse band on the autoradiograph. This loss of sharpness is kept to a minimum both by drying the gels (to reduce the depth of the source) and by using a low energy source such as ^{35}S. In fact ^{35}S is so weak a source that drying of the gels is essential to prevent severe quenching of the signal.

5.5.1 Drying and autoradiography of the gel

(i)   At the end of electrophoresis, disconnect the power from the gel and discard the TBE from the buffer chambers. Remember that the lower buffer will be contaminated with both unincorporated radiolabel and radioactively-labelled short DNA molecules and should be disposed of appropriately.

(ii)   Remove the chambers and the sealing tape from the glass plates.

(iii)   Insert a thin spatula between the plates, notch upward, and very gently separate them by twisting the spatula. If all goes to plan, the gel should adhere to the larger non-silanized plate underneath; if it sticks to the notched plate it has to be remembered that the clone and track order of the resulting autoradiograph will be reversed. If the gel sticks to both of the glass plates it is best to separate them submerged within the fixing solution.

(iv)   Gently lower the plate and gel into a tray containing sufficient 10% acetic acid in water to cover it (if the gel can be kept stuck to the plate, fixing/urea dialysis will take longer, but troublesome ripples in the gel will be avoided) and leave it for at least 15 min. Leaving the gel for too short a time in the fixing solution does not dialyse out all the urea which makes drying of the gel more difficult.

(v)   Slowly remove the glass plate and gel from the acetic acid and let as much as possible of the liquid drain off. A sheet of plastic mesh, such as greenhouse shading, and a second plate can be used to keep the gel in position during removal from the fixing bath.

(vi)   In one smooth movement, lay a piece of 3 MM chromatography paper, cut to a suitable size, on top of the drained gel and press down gently to get good gel to paper contact. Starting at one corner at the well end of the gel, peel back the paper, and the gel should adhere securely to it. If it looks as though the gel has not stuck well to the paper (which sometimes happens with high percentage acrylamide gels), place another glass plate on top of the paper to enable flipping the whole assembly over. The top glass plate can then be slowly lifted away allowing the gel to fall down slowly onto the paper.

(vii)   Put a layer of Saran plastic wrap on top of the gel and trim the paper/gel/plastic wrap to remove the excess and position it, paper side down on a gel drier.

(viii)   Dry the gel under vacuum at 80°C for at least 15 min. The time taken for the gel to dry can be much longer than this if the vacuum is poor but it is essential that it is completely dry to avoid it sticking to the X-ray film.

(ix)   Peel away the plastic wrap and put the dry gel in direct contact with a sheet of film, inside a film cassette, making sure the correct surface of the dry gel and film (if single sided film is used) are in contact, and expose for 18−48 h.

(x)   Develop, fix and dry the film according to the supplier's recommendations.

5.6 Multiple loading

The size and format of polyacrylamide gels can be anything you like. They are normally based upon X-ray film and film cassette sizes, either 20 cm or 40 cm wide and 40 cm upward (even to 2 m) long. A 40 cm long 6% acrylamide sequencing gel will give up to 200 nucleotides of data if resolution is good, and will take approximately 2 h to run. The run time will depend upon the position of primer hybridization, relative to the cloning site used. Using the universal primer entails sequencing through part of the vector polylinker and the first few nucleotides beyond the priming site will not normally be detectable anyway. The data yield obviously depends upon sequence quality and gel resolution. Under normal experimental conditions, where the occasional sequence and gel does not approach the ideal quality, the average over several hundred gels can drop to as low as 150 nucleotides per sequence reaction. By increasing the length of the fractionating gel to 50 cm the run time increases proportionally (to 2.5 h) as does the amount of data gained (up to around 250 nucleotides). By the time the smaller DNA chains reach the bottom of the longer gel, the larger chains have been fractionated through a greater length of gel, the separation between the bands has increased and interpretation over a greater length is made possible. Unfortunately, extrapolating this tactic to longer and longer gels to give longer sequences does not work. For increasing gel length, the run time continues to rise proportionally but the amount of sequence data gained fails to keep up. The practicalities of very long gels soon comes into question also. The glass plates become cumbersome and difficult to handle safely, pouring them bubble free is more difficult and gel drying and autoradiography become increasingly complicated. A preferred approach is to load the same sample at time intervals either on the same gel or by running two or three gels for different times. The data run off the bottom of the long run is obtained from the shorter run. Two loadings on a 40 cm long 6% acrylamide gel, one run 2 h and one run 4 h, will give in total around 300 nucleotides with an overlap between the two runs. When separate gels are used for the different run times, long runs can be electrophoresed more quickly by reducing the total acrylamide concentration in the gel, typically to around 4%. Lower percentage gels have reduced mechanical strength, however, and can be very difficult to handle.

Having attempted these ways of maximizing the sequence data obtained from each template sequenced, it is then found that the maximum of sequence data readable is limited by the resolving power of polyacrylamide gels. Although bands are separated from each other and patterns are clearly discernible, a determination of their precise order is not possible. Claims are often made that lengths of 600 nucleotides or even more can be read; we would venture that this is under exceptional circumstances and far from routinely practicable. As the interband separation decreases the accuracy of the interpreted sequence falls markedly. Most often this is because either the number in a 'run' of multiple bands is difficult to determine, as the grains on the film tend to merge, or consecutive bands in separated tracks are read transposed, or the mobility problems associated with compressions have been undetected and bands missed. An inaccurate sequence is of little more value than no sequence at all. Under normal conditions, difficulty of interpretation and reduced quality will limit the maximum sequence length which can be read to around 400−450 nucleotides from three multiple loadings on 40-cm gels.

Now if the total sequence that was required was 450 bases and it proved possible to read it unambiguously in both orientations from two templates, then this approach would be perfectly satisfactory. It is a far more likely situation, however, that this would be only a small subfragment of a much larger sequencing project. If we analyse what would have been achieved, using the multiple loading tactic, even under the most favourable of circumstances, it can be seen that the data output of the gels has been reduced from an average of 200 bases per gel to 150 bases per gel (three loadings giving a total of 450). If these three gels had been used to determine the sequence of different templates, a total of around 600 bases could easily have been achieved. Using the random strategy, the gels are currently the rate limiting step to rapid sequencing and it is important to maximize the yield obtained from each as far as is possible.

5.7 Field strength gradient gels

It has been stated above that using currently available techniques, there is a maximum length we can realistically aim for in the fractionation of sequencing reaction products on polyacrylamide gels. This maximum can only be obtained by electrophoresis for extended periods (around 6−8 h) on low percentage acrylamide gels. If the shorter molecules are not to be run off the bottom and the data lost, this can only be gained by employing very long gels or by running time staggered loadings ensuring an overlap of data between them. An effective alternative is to modify the uniform electric field within the gel so that there is a gradient of voltage drop per unit length across it, high at the top and low at the bottom. This results in a gradient of driving force over the gel length. DNA molecules within the gel experience a larger migrating force than those immediately below them, the smallest and fastest moving molecules always experiencing the least driving force. The overall effect of such a system is that the small DNA molecules slow down towards the bottom of the gel, tending not to be electrophoresed off and thereby reducing the interband spacing whereas the larger chains are fractionated through a greater length of gel, just as achieved in long runs. The field strength gradient is commonly produced by varying either the thickness of the gel (5) or the ionic concentration of the gel across its length (4). A comparison of these techniques used in sequencing gels is shown in *Figure 9*.

5.7.1 *Wedge gels*

The electrical resistance of a conductor changes with cross-sectional area. That of TBE/polyacrylamide gels decreases with increasing cross-sectional area. A uniform thickness gel has an essentially equal voltage drop per unit length from top to bottom. If the gel thickness is increased linearly down its length, the voltage drop per unit length decreases linearly in accordance with Ohm's law. Such a thickness gradient is simple to produce, either by using commercially available wedge shaped side spacers or by fabricating your own by glueing short pieces of side spacer onto the bottom of normal ones. The slope of the gradient can be increased by adding on more and more pieces. Unfortunately each doubling of gel thickness doubles the drying time needed and also reduces the sharpness of the lower bands significantly. Once a thickness approaching 1.5 mm is reached both of these become unacceptable. This technique was used very effectively to increase the average number of nucleotides read from a 40-cm gel to around 250. The use of wedge gels led to the introduction of ionic strength gradient gels which removed all the drawbacks of the wedge gel and superseded them. However, their ease of construction and pouring still makes 'wedges' a popular alternative.

5.7.2 *Preparing a wedge gel*

(i)   Cut four 3-cm lengths of Plastikard side spacer and using polystyrene fluid cement fix one on each side of two normal spacers at one end.

(ii) Assemble the gel mould, as in Section 5.2.1 above, with the thicker end of spacer at the bottom.

(iii) Pour the gel using the method in Section 5.2.1 remembering that more gel mix will be required. Clamp the plates with foldback clips, at the top and over the thick part of spacer at the bottom.

(iv) Use the gel running conditions described in Section 5.4.1 except that the running time needed will increase to around 2.5 h for a 40-cm gel or around 3.5 h for a 50-cm gel.

(v)   After running the gel, allow longer fixing and drying times.

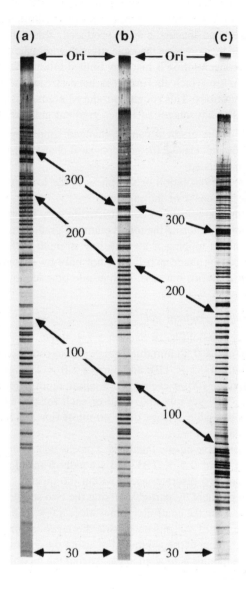

Figure 9. A comparison of the band separation of linear, wedge and gradient gels. The T reaction of a sequencing experiment carried out on M13 mp8 DNA was run on three different 40 cm long gels, (**a**) a standard linear 6% polyacrylamide gel, (**b**) a 0.35−1.05 mm wedge shaped 6% polyacrylamide gel and (**c**) a 0.5×−5.0×TBE buffer gradient 6% polyacrylamide gel. The position of some fragment lengths (taken from the 5′ end of the primer) are marked for ease of comparison. The increased separation from the origin (ori.) of the longer fragments on both the wedge and buffer gels increases the amount of readable data.

5.7.3 *Buffer gradient gels*

Buffer gradient gels were introduced in response to the shortcomings of wedge gels (4). A more even interband spacing is again produced, although this time the varying electric field is effected by increasing the ionic strength within the gel toward the bottom. A gradient of buffer concentration is made by limited mixing, within a pipette, of two different gel solutions (to which the initiators have already been added) containing different TBE concentrations. This roughly produced gradient is then pipetted directly into the gel mould. The advantages of buffer gradient gels are:

(i)   The steepness of the gradient can be changed simply by altering the strength of TBE, this does not change the gel's physical characteristics and does not alter the drying time needed.

(ii)   There is no loss of resolution or sharpness in the lower part of the gel despite increasing the steepness of the gradient.

(iii)   The field strength gradient can be restricted to only a part of the lower portion of the gel simply by altering the total volume of gradient mix and after pipetting this into the mould topping it up with low strength TBE gel mix.

(iv)   The gradient can be made non-linear, not only by changing the total volume occupied by it as in (iii) but also, with a little practice, by changing the degree of mixing.

5.7.4 *Preparing a buffer gradient gel*

(i)   Assemble the gel mould as in Section 5.2.1 above.

(ii)   For a 20 × 50 cm × 0.35 mm thick gel, warm to room temperature in separate containers 50 ml of 0.5 × TBE and 7 ml of 5.0 × TBE gel mix (see *Table 8*).

(iii)   Add, with mixing, 100 μl each of 25% ammonium persulphate and TEMED for the 0.5 × TBE gel mix and 14 μl of each for the 5.0 × TBE gel mix.

(iv)   Into a 50-ml disposable syringe draw 40 ml of 0.5 × TBE gel mix and put it aside for use later.

(v)   Into a 10-ml disposable pipette fitted with a pipette controller (such as a Propipette) take up first 6 ml of 0.5 × TBE and secondly 6 ml of 5.0 × TBE gel mix. These will form two discrete layers within the pipette.

(vi)   Make a rough gradient by partially mixing the two solutions. This can be done very simply by using the controller to introduce a few air bubbles. These bubbles will rise to the top of the pipette mixing the layers as they pass the interface.

(vii)   Transfer this gradient to the gel mould by using the controller to pipette it down one edge of the mould, held at an angle, so that the mixture flows smoothly and evenly to the bottom. More even gradients can be produced by pouring down the centre of the mould. This takes a bit more practice and demands very clean, polished glass plates if air pockets are to be avoided.

(viii) Once the pipette is empty, immediately lower the mould to horizontal which will stop the flow of gel mix.

(ix)   Pick up the syringe containing the 0.5 × TBE gel mix and continue to fill the mould by raising the top edge to regain a good pouring angle. The flow of the two gel mixes should again be kept continuous and should be directed to wash the gradient mix toward the bottom. The rate of flow can be quite accurately controlled by altering the angle at which the mould is held.

(x)   When completely filled, lower the mould to close to horizontal and insert the slot former.

(xi)   Clamp the plates together, over the side spacers, with foldback clips (two or three over each edge) to ensure an even thickness of gel, and leave it for 30 min to polymerize.

(xii)   Use the gel running conditions described in Section 5.4.1 except that the running time needed will increase to around 2.5 h for a 40-cm gel or around 3.5 h for a 50-cm gel.

6. REFERENCES

1. Sanger,F., Nicklen,S. and Coulson,A.R. (1977) *Proc. Natl. Acad. Sci. USA,* **74**, 5463.
2. Bankier,A.T., Weston,K.M. and Barrell,B.G. (1987) In *Methods in Enzymology.* Wu,R. (ed.), Academic Press, New York, Vol. 155, p. 51.
3. Sanger,F. and Coulson,A.R. (1978) *FEBS Lett.,* **87**, 107.
4. Biggin,M.D., Gibson,T.J. and Hong,G.F. (1983) *Proc. Natl. Acad. Sci. USA,* **80**, 3963.
5. Bankier,A.T. and Barrell,B.G. (1983) In *Techniques in the Life Sciences.* Flavell,R.A. (ed.), Elsevier Scientific Publishers, Ireland, Vol. B5, p. 1.
6. Smith,L.M., Sanders,J.Z., Kaiser,R.J., Hughes,P., Connell,C.R., Heiner,C., Kent,S.B.H. and Hood,L.E. (1986) *Nature,* **321**, 674.
7. Prober,J.M., Trainor,G.L., Dam,R.J., Hobbs,F.W., Robertson,C.W., Zagursky,R.J., Cocuzza,A.J., Jensen,M.A. and Baumeister,K. (1987) *Science,* **238**, 336.
8. Hong,G.F. (1982) *J. Mol. Biol.,* **158**, 539.
9. Poncz,M., Solowiejczyk,D., Ballantine,M., Schwartz,E. and Surrey,S. (1982) *Proc. Natl. Acad. Sci. USA,* **79**, 4298.
10. Barnes,W.M. and Bevan,M. (1983) *Nucleic Acids Res.,* **11**, 349.
11. Henikoff,S. (1984) *Gene,* **28**, 351.
12. Dale,R.M.K., McClure,B.A. and Houchins,J.P. (1985) *Plasmid,* **13**, 31.
13. Deininger,P.L. (1983) *Anal. Biochem.,* **129**, 216.
14. Anderson,S. (1981) *Nucleic Acids Res.,* **9**, 3015.
15. Eperon,I.C. (1986) *Anal. Biochem.,* **156**, 406.
16. Kristensen,T., Voss,H. and Ansorge,W. (1987) *Nucleic Acids Res.,* **15**, 5507.
17. Mills,D.R. and Kramer,F.R. (1979) *Proc. Natl. Acad. Sci. USA,* **76**, 2232.
18. Mizusawa,A., Nishimura,S. and Seela,F. (1986) *Nucleic Acids Res.,* **14**, 1319.
19. Tabor,S. and Richardson,C.C. (1987) *Proc. Natl. Acad. Sci. USA,* **84**, 4767.
20. Peterson,M.G. (1988) *Nucleic Acids Res.,* **16**, 10915.
21. Innis,M.A., Myambo,K.B., Gelfand,D.H. and Brow,M.D. (1988) *Proc. Natl. Acad. Sci. USA,* **85**, 9436.
22. Martin,W.J. and Davies,R.W. (1986) *Biotechnology,* **4**, 890.
23. Wada,A. (1987) *Nature,* **325**, 771.

7. APPENDIX

7.1 **Materials and equipment**

(i)   Microcentrifuge: MSE Micro Centaur.
(ii)   Multi tube vortexer: SMI Model 2601.
(iii)   GF/C filters (7 mm), Whatman International Ltd. Cut filters from larger sheets with a cork borer.
(iv)   Sintered glass filtration unit and water pump.
(v)   Gel apparatus: Model EV200 (20 × 20 cm), EV400 (20 × 40 cm) or EV500 (20 × 50 cm), Cambridge Electrophoresis Company, 84 High Street, Cherry Hinton, Cambridge CB1 4HZ, UK. With temperature control, the Pharmacia-LKB Macrophor.
(vi)   Polyester gel tape, Cat. No. 49LY, Neil Turner Ltd, Watlington, nr. Kings Lynn, Norfolk, UK.
(vii)   Plastikard 0.35 mm × 44 cm × 66 cm (for making spacers and combs), Slaters (Plastikard) Ltd, Temple Road, Matlock Bath, Matlock, Derbyshire DE4 3PQ, UK.
(viii) Power supplies: Pharmacia-LKB Model 2197 2.5 kV, 200 mA, with constant power capability.
(ix)   X-ray film: Fuji RX, Kodak XAR and X-ray film cassettes with snap lock 35 × 43 cm.
(x)   Microtitre trays: Becton Dickinson-Falcon flexible assay plates No. 3911 round bottomed 96-well.
(xi)   Hamilton PB600-1 repetitive dispenser fitted with 1710LT 100 μl gas-tight syringe with Luer tip and adaptor, Cat. No. 31330.

(xii) Damon IEC Centra 4X bench top centrifuge fitted with microtest plate head.

(xiii) Dow Corning Saran Wrap clingflim.

(xiv) Water bath or heat block for Sequenase reactions (65°C, 35°C 37°C, 80°C).

(xv) Ovens for Klenow reactions (55°C, 80°C), incubating plates (37°C).

(xvi) Environmental incubator shaker: New Brunswick Model G24.

(xvii) Gilson Pipetman P20, P200, P1000.

(xviii) Drummond Sequencing pipette: Drummond Scientific Company, 550 Parkway, Broomhall, Pennsylvania.

(xix) Gel drier: Bio-Rad Model SE 1125B and oil vacuum pump with a low temperature trap.

(xx) Robot pipettor: Beckman Biomek 1000 using MP20 tool.

CHAPTER 3

Troubleshooting in chain-termination DNA sequencing

E.SALLY WARD and CHRISTOPHER J.HOWE

1. INTRODUCTION

Although dideoxy sequencing is regarded by many as the method of choice for any large sequencing project, it is often found to be a rather more difficult technique than chemical sequencing to set up afresh in a laboratory. This chapter is intended as a guide to the problems most frequently encountered, and should be consulted in conjunction with the two preceding chapters. It is aimed at the laboratory which has expertise in the elementary aspects of handling nucleic acids, and therefore does not cover the problems associated with routine operations, for which the reader is instead recommended to consult the many molecular biology recipe manuals available (e.g. 1–3). Many of the problems discussed here will be seen as completely trivial to an experienced sequencer—we make no apology for including them, as this chapter is aimed at the beginner for whom even the most trivial problems can appear insurmountable. Problems specifically associated with sequencing of double-stranded DNA are covered in Chapter 4.

2. PROBLEMS ENCOUNTERED IN CLONING

2.1 Generating recombinants for transfection

The first indication that problems are occurring in the cloning of DNA (usually specific restriction fragments, end-repaired fragments produced by sonication or DNase digestion, or material produced by exonucleolytic degradation of larger fragments) into M13 generally arises after the transfection of host cells with supposedly recombinant M13-insert molecules. It is therefore important to include suitable controls in the transfection. The more informative transfections are shown in *Figure 1*, together with the expected and idealized results on plates containing Xgal and IPTG. Of these, transfections (i), (ii) and (vi) are the most important. Deviations from these patterns indicate various problems, and are discussed below, in order of decreasing frequency.

2.1.1 *Transfection (i) gives few white plaques—other results as expected*

This indicates that the vector-insert ligation has failed to take place, although the vector can self-ligate prior to phosphatase treatment, since (iv) gives many blue plaques. One of the most likely causes is that the molar ratio of vector DNA to insert DNA may be suboptimal. This can be identified and rectified by repeating the ligations at a range of vector-insert ratios.

Expt. No.	Vector cut with RE?	Vector phosphatased?	Ligase added?	Insert added?	Expected Blues/Whites	Idealized Blues/Whites
(i)	Yes	Yes	Yes	Yes	Few/Many	None/Many
(ii)	Yes	Yes	Yes	No	Few/Few	None/None
(iii)	Yes	Yes	No	No	Few/V.few	None/None
(iv)	Yes	No	Yes	No	Many/Few	Many/None
(v)	Yes	No	No	No	Few/None	None/None
(vi)	No	No	No	No	V.many/V.few	V.many/None

Figure 1. Transfections useful in M13 cloning. The significance of the reactions is explained in the text.

   Another likely cause is that the insert material is unsuitable, either because it contains ligase inhibitors (which may be present as a result of extraction of DNA from gels) or has unsuitable ends (which may be due to inefficient end repairing, or nuclease contamination of the restriction enzymes used to generate it). The presence of ligase inhibitors can be demonstrated by repeating (iv) with insert added, resulting in inhibition of vector self-ligation and a dramatic fall in the number of blue plaques. If this is consistently found to be the problem, a different source of gel material or method of fragment isolation should be employed. Note that certain agarose preparations contain rather efficient inhibitors of ligation that may be extracted from the gel with DNA, probably including free sulphated polysaccharides (1) so this may be a problem if insert material has been isolated using low gelling temperature agarose (unless the latter is known to be of good quality). Other ways of isolating DNA from gels include electrophoresis onto DEAE-cellulose paper (4), NA45 membrane (5), or dialysis membrane (1,6), and electrophoresis into wells cut in the gel (1). Alternatively, further purification of the extracted DNA may be attempted, for example using 'Elu-tips' (3) or Spin-X tubes (see Chapter 1, Section 3.3.15). If the insert material has unsuitable ends, end repairing should be repeated, with altered incubation times or amounts of polymerase if necessary, or alternative restriction enzyme preparations should be tried.

   The problem can on occasions be caused by contamination of the phosphatase with nuclease activity, or failure to purify the vector fully away from phosphatase. Provided the phosphatase is from a reputable source and has been used and stored carefully, nuclease contamination is not usually a problem, although there is often significant difference in nuclease activity between batches. Repeated heat inactivation of phosphatase and phenol extraction should remove phosphatase contamination from the vector. The greater the white:blue ratio in (i) than the ratio in (ii), the safer the plaques in (i) are to use.

2.1.2 *Many blue plaques in (i) and (ii)—other results as expected*

This indicates that the vector DNA has not been adequately phosphatased, and therefore self-ligation is taking place. Phosphatase preparations as supplied (e.g. lyophilized or in glycerol) are generally stable at 4°C. Once diluted, storage is possible at −20°C, but not advisable. The white plaques in (i) are safe to use.

2.1.3 *Transfection (i) gives few white plaques, (iv) gives few blue plaques—other results as expected*

This indicates that the ligase has low activity, either because of defective enzyme/buffer or because of inhibitors in the vector DNA. Replacement of the enzyme, buffer and vector DNA should indicate the problem. It is inadvisable to store ligase buffers for periods much greater than a week if they include DTT or ATP. White plaques in (i) are probably safe to use.

2.1.4 *Many blue plaques in (i), (ii), (iii) and (v)—other results as expected*

This indicates that the restriction enzyme has failed to digest the vector, leaving closed circular material which will give blue plaques regardless of subsequent phosphatasing. Digestion of vector should be checked by running an agarose gel, but note that quantities of undigested DNA too small to be readily visualized in a gel can give a very high background of blue plaques, because transfection with undigested vector is much more efficient than digestion, religation and transformation. The white plaques in (i) are safe to use.

2.1.5 *No plaques at all*

This suggests a very low transfection efficiency. See Section 2.2.

2.1.6 *Transfection (ii) gives many white plaques—possibly also (iv)*

This indicates that DNA other than the desired insert is being ligated into the vector. It can be caused by contamination of vector, ligase, phosphatase or buffers with low levels of nuclease (fragmenting some vector DNA, which is then ligated into intact vector) or (less likely) DNA. Systematic replacement of each of these components should indicate which is at fault. White plaques in (i) are NOT safe to use.

2.1.7 *Many white plaques in (vi), and probably others*

This indicates contamination of the vector strain (see Section 2.2.2). A blue plaque picked from (vi) may be safe to grow up more vector DNA. White plaques in (i) are NOT safe to use.

2.1.8 *No blue plaques at all, but many white ones*

This suggests that the IPTG or Xgal have been omitted or become inactive, or that the colour has simply failed to develop. This can be caused by incubation at too high a temperature. Leaving plates out on the bench or at 4°C for a few hours may enhance blue colour. The wrong host strain may have been used (see Section 2.2). White plaques in (i) are probably unsafe to use.

2.1.9 *Very few plaques in (i) − (v); (vi) as expected*

This is likely to be caused by severe nuclease contamination. Integrity of DNA can be assessed by gel electrophoresis (although this may not reveal low levels of exonucleolytic activity). Demonstration of endonucleolytic activity may also be possible by incubation of uncut DNA with buffers or ligase, followed by transformation, when a significant reduction in the number of blue plaques would be expected. Any white plaques in (i) may not be safe to use.

2.2 **Problems in transfection**

2.2.1 *Standard microbiological technique*

Some of the problems encountered here are 'trivial' ones associated with pouring suitable lawns and so on. The most common ones are:

(i) *Large clear areas up to 1 cm in diameter.* These are usually caused by condensation, generally resulting from inadequate drying of plates before pouring lawns. Plates should also be incubated with the agar side uppermost (ensuring that the soft agar has set before inverting the plates). Clear areas may be plaques caused by contaminating bacteriophages, many of which will give much larger and clearer plaques than M13. Suspect media should be autoclaved and discarded, and scrupulous attention paid to sterilizing equipment and media before use.

(ii) *Streaking of plaques.* Instead of being circular, plaques have a 'comet-shaped' appearance. This is also usually due to inadequate drying of plates.

(iii) *Mottling of lawns.* The lawn has a very uneven appearance, which makes it impossible to distinguish plaques. This is usually caused by partial solidification of the top agar before pouring.

2.2.2 *Transfection*

(i) *A very thin lawn, with no plaques visible.* This usually indicates more or less confluent lysis. It is unlikely that all the transfections outlined above will lead to confluent lysis, and a more probable explanation is that the media or the host culture used are contaminated with another phage, not necessarily M13. Fresh media should be made up, and a new host culture set up. Contaminated media can be identified if necessary by spotting a few microlitres of suspect material on a freshly seeded lawn, allowing to dry by leaving the plate open near a lit bunsen burner and then incubating.

Another cause of a very thin lawn can be top agar which is too hot. If the vessel containing the molten top agar cannot be comfortably held in an unprotected hand, the agar is almost certainly too hot, and will kill most of the cells before plating. Top agar should remain molten in a 42°C incubator or waterbath, at which temperature it should be quite safe to use.

(ii) *An apparently normal lawn, with no plaques visible.* It is unlikely that the cells will fail to become competent, so before discarding such plates, check that there are indeed no plaques. If the lawn was seeded too thickly (i.e. too many cells used for each transformation) then the plaques may be too small to be readily distinguished without careful examination. Tiny plaques may also rise from the insertion of too large a piece of DNA, slowing down phage replication and also increasing the selective advantage of any deletion mutants that arise. Furthermore, deletion mutants may themselves give small plaques.

It seems that a lack of plaques can also be caused by use of top agar which is too hot (albeit not hot enough to kill the cells as stated above). Possibly this causes some physiological shock, which inhibits the growth of phage in the host.

The effect can also be caused by a loss of the F' plasmid from the host. This results in a loss of the sex pilus, which is required for the phage to infect cells. Continued subculturing of the host strain in a rich broth or on rich plates is the commonest cause of this, as there is consequently no selection for retention of the plasmid. Cultures of the host strain should therefore be set up by inoculating broth (usually YT) with a single colony from a minimal plate (which therefore lacks proline). The ability to synthesize proline depends on genes on the F' plasmid for all hosts used in sequencing work, so selection for proline prototrophy ensures the presence of the plasmid. Genotypes of the two strains most commonly used and a recipe for F' selection are given in Section 2.2.3.

2.2.3 *Strain genotypes and proline selection*

JM101 (7) K12, Δ(*lac pro*) *supEthi*-1F'*traD36proA*$^+$B$^+$*lacI*^{q},ZΔM15

TG1 (8) K12, Δ(*lac pro*)*supEthi*-1*hsd*D5 F'*traD36proA*$^+$B$^+$*lacI*^{q},ZΔM15

Minimal medium: 2% agar in distilled water, autoclave in 300-ml quantities, add 100 ml of minimal salts to 300 ml of molten agar, carbon source (e.g. glucose) to 0.4% w/v, thiamine to 1 mg 1^{-1}. Minimal salts: K_2HPO_4 28 g, KH_2PO_4 8 g, $(NH_4)_2SO_4$ 4 g, $(Na)_3$citrate 1 g, $MgSO_4$ 0.4 g, distilled water to 1 litre, autoclave in 100-ml quantities.

3. PREPARATION OF SINGLE-STRANDED DNA

Most of the problems associated with this set of operations do not become apparent until the sequencing gels themselves have been run and the autoradiographs developed. A few points may be worth making, however. Generally, the pellet produced by NaCl/PEG precipitation of the M13 phage should be fairly readily visible, and about the size of a pinhead. If the pellet is much larger, it is possible that the cultures were incubated for too long during growth of the phage, or that the cell density of the starting culture was too great. This is likely to lead to degradation of the DNA, (as a consequence of cell lysis) and therefore a high level of artefact bands in the sequencing gels – see Section 4.1.3 and *Figure 2*.

A number of factors may cause the pellet to be very small, or even invisible. The most obvious is a failure to inoculate the culture with sufficient viable phage. This may happen if the plate containing the plaques has been kept for too long before using them to prepare single-stranded DNA. The 'life expectancy' of plaques on a plate is variable, but ideally one should aim to use them within 24 h of their generation. If they are kept for longer, storage in a refrigerator is necessary. Deletion of the insert (see Sections 2.2.2 and 4.1.2) is also more likely the longer the plaques are kept.

Low yield may also be due to having too low a cell density at the time of inoculation with phage, or using an unsuitable host strain. Aside from the possibility of contamination with strains that do not support growth of M13, this problem may again be caused by loss of the F' plasmid, resulting in fewer cells in the culture being infected by phage and thus giving a lower titre. This problem can be circumvented by use of colonies picked off a selection plate which does not contain proline (see Section 2.2.2 and 2.2.3).

High or low growth temperatures may also inhibit phage growth, so the incubator

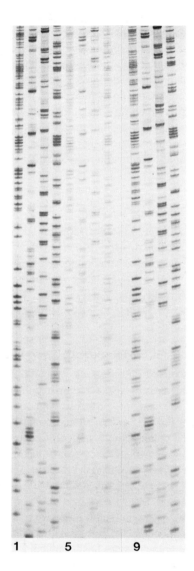

Figure 2. Effect of growth conditions on ssDNA template preparation. **Lanes 1−4**; T,C,G,A tracks of template isolated from cells grown for twice the normal length of time (10 h), after infection of an *Escherichia coli* culture of four times the usual cell density. **Lanes 5−8**; T,C,G,A tracks of template isolated from *E. coli* cells after infection of a culture of four times the usual cell density, and grown for 5 h post infection. **Lanes 9−12**; T,C,G,A tracks of template isolated from *E. coli* cells grown for twice the normal length of time after infection.

should be checked. Good aeration of the cultures is also important for phage infection and growth, so they should be shaken at 200−250 r.p.m. during growth. Even if the phage pellet is scarcely visible at all, however, it is worth continuing with the DNA preparation, although it may be helpful to dissolve the DNA produced at the end in half the usual volume of buffer, to avoid having a very dilute solution for sequencing (which would then require prolonged autoradiography). Note that, even with a normal sized phage pellet, one should not expect to see a *nucleic acid* pellet at the end of the preparation, although some 'ghosting' may be apparent at the bottom of the centrifuge tube. A clearly visible pellet usually indicates contamination of the DNA, most likely with salt (i.e. the sodium acetate used in precipitation) or chromosomal DNA, but possibly also PEG or phenol, and will usually result in poor quality sequence (see Section 4.1.3). It may be possible to rescue this by redissolving and then reprecipitating the DNA before sequencing, but it is sensible to try sequencing with the DNA first to determine the seriousness of the problem.

4. PROBLEMS ENCOUNTERED WITH SEQUENCING REACTIONS AND GELS

After preparing the single-stranded DNA, the next stage at which problems are likely to manifest themselves is when the autoradiograph of the sequencing gel has been developed, and the sequence is found to be too faint, the bands are too diffuse, or there are bands in more than one track at the same level. Some of these problems cannot be easily rectified, and allowance must then be made in reading the gel. Guidance on this is given in Section 5. Many problems can be fairly easily solved however, and these are discussed below, with photographs of specimen gels illustrating some of them. Defects are grouped according to whether they are caused by deficiencies in template, priming, sequencing reactions or electrophoresis. Note however that different deficiencies can give rise to the same overall appearance.

4.1 Poor quality template

This is probably one of the most common areas of difficulty, and one can often expect one's very first template preparations to be of rather low quality, without any one particular fault predominanting. With practice, a rapid improvement is usually seen. It should also be mentioned that the quality of template needed is often determined by the DNA polymerase used. Some preparations will give perfectly adequate sequence from even quite low quality DNA, which might give totally unreadable results with other enzyme preparations. In our experience, 'Sequenase' (Chapter 2, Section 3.3) is less affected by poor quality DNA preparations.

4.1.1 *Sequence of good quality but very faint*

This is likely to be due simply to a low yield in the DNA preparation (see Section 3). A longer exposure in autoradiography may be sufficient to compensate, but this will slow down a sequencing project of any size.

Contamination of DNA pellets with salt causes a dramatic inhibition of polymerase, resulting in very faint gels. The sequence often also has artefact bands (see below).

4.1.2 *No sequence visible at all*

This can be due to a complete failure of the phage to grow (in which case no phage pellet would be seen)—see previous paragraph and Section 3. Alternatively, deletion of part of the phage genome may generate mutants which give white plaques on Xgal plates, but to whose DNA the primer cannot anneal. Such deletion mutants may often outgrow the other phage. They are more likely to be a problem on plates which have been stored for some time after transfection.

4.1.3 *Sequence shows a high occurrence of artefact bands (i.e. bands at the same level in more than one track)*

Contamination with salt and/or PEG are often blamed for generating artefact bands, and also making bands more diffuse. In our hands, PEG at least is not usually a major problem (indeed deliberately adding PEG to the DNA preparations had no significant effect), but this may be a reflection of the enzyme preparation used (see Section 4.3.1). Increasing the amount of enzyme added may help to reduce the number of artefact bands. Note that a high salt concentration will allow a less stringent annealing between primer and template. This may be particularly problematic if the template is contaminated with chromosomal DNA.

Artefact bands can also be caused by nicking of the template DNA during preparation, often as a result of incubating the cultures for too long before harvesting the phage, or having too high a cell density at the start of the incubation (*Figure 2*). It is probable that some cell lysis occurs, liberating nucleases which attack the DNA. RNA fragments may also be released, and act as random primers in sequencing.

Occasionally, artefact bands throughout a sequence may be generated as a result of careless picking of plaques for phage growth, resulting in picking a mixture of two different phage. This will obviously result in the superposition of two sequences (not necessarily of equal intensity) throughout, but should be found only in isolated DNA preparations within one batch. It is more likely to be a problem when plates have not been properly dried and condensation spreads phage particles over the surface of the lawn, when the plaque density is very high, or when plates have been stored for several days before using the plaques, since phage particles can diffuse through the soft agar.

Single-stranded DNA preparations should be kept frozen, when they are generally quite stable. Repeated freeze-thawing should not usually be necessary, as any one clone should not need to be sequenced more than a few times at most. Although a few cycles of freeze-thawing do not usually lead to marked deterioration of sequence quality (notably the appearance of artefact bands combined with a general reduction in band intensity), this should be avoided as far as possible. If it is necessary to make fresh template from a single-stranded DNA stock, this can be done by transfection in the usual way, even though the DNA is single-stranded. Generally 1 μl of a 100 $\times$ diluted stock of sequencing template gives a suitable number of plaques on the lawn. Note that a small proportion of molecules in the stock may contain deletions or other rearrangements, so it is advisable to work up several plaques from the retransfected lawn.

4.2 **Priming**

4.2.1 *Sequence of good quality but faint*

This can be caused by using too low a concentration of primer, or by carrying out the annealing at much too low or high a temperature (*Figure 3*). Repeated freeze-thawing of the primer will bring about its degradation, which will also make the sequence rather faint. In addition, non-specific annealing of primer fragments may increase the

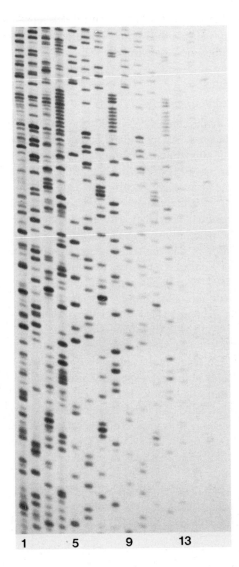

Figure 3. Effect of primer concentration on sequencing reactions. **Lanes 1−4**; T,C,G,A tracks of template using 15 picomoles of primer per track. **Lanes 5−8**; T,C,G,A, tracks of template using 0.05 picomoles of primer per track. **Lanes 9−12**; T,C,G,A tracks of template using 0.005 picomoles of primer per track. For **lanes 1−12**, primer was annealed to template at 55−60°C. **Lanes 13−16**; T,C,G,A tracks of 0.05 picomoles of primer per track annealed to template at 20°C.

background of artefact bands. Stock solutions of primer should be kept in small aliquots at −20°C (or below) to avoid excessive freezing and thawing.

4.2.2 *Sequence has many artefact bands, but is not faint (may be abnormally strong)*

This can be caused by having too high a concentration of primer. The artefact bands generated often have a rather uneven spacing, by comparison with 'normal' artefact bands. A titration of various primer concentrations with representative template DNA preparations will usualy indicate the optimal quantity to use, and should normally be carried out when starting a new batch of primer. See *Figure 3*.

4.3 Sequencing reactions

4.3.1 *Sequence has many artefact bands*

Assuming the quality of the template is good, the most likely cause of this is the polymerase preparation used. As mentioned earlier, some polymerase preparations appear more tolerant of suboptimal conditions than others, and increasing the concentration of polymerase may be helpful. Anything which is likely to decrease the activity of the polymerase should be avoided. Factors which are important here include keeping the polymerase stock at $-20°C$ (but not at $-80°C$, when the repeated freeze-thawing needed will rapidly denature the enzyme), avoiding diluting the enzyme until just before it is to be added to the reaction mixes (regardless of whether the diluted enzyme is kept on ice before addition), and the temperature at which the reactions are carried out (although we see little difference between reactions carried out at room temperature and those carried out at $37°C$, except when templates have a high degree of secondary structure). Of particular importance, however, seems to be the quality of dithiothreitol added to the sequencing reactions. This is especially so when (as is now usual) sequencing is carried out with $[\alpha\text{-}^{35}S]dATP$ rather than $[\alpha\text{-}^{32}P]dATP$. DTT solutions are not very stable, even at $-20°C$, a fact which is exacerbated by repeated freeze-thawing. Deterioration of DTT can result in the appearance of artefact bands and reduction of intensity of genuine bands in the gel, especially pronounced in regions nearer the top of the gel. Different tracks may show this to greater extents—very often the C track is one of the first to be affected. Although it is not necessary to make up a fresh DTT solution from solid every day, it is wise to do so each week, and divide it into aliquots to avoid freezing and thawing. Solutions of DTT have a characteristic smell, and any diminution of this smell usually indicates deterioration of the solution, which should be discarded.

Artefact bands restricted to one track can be due to an unsuitable nucleotide mix (see below). This track will usually be significantly fainter than the others. Cross-contamination of mixes will also, obviously, lead to artefact bands.

4.3.2 *Uneven distribution of radioactivity throughout the gel*

It is sometimes the case that the bands towards the bottom of the gel are much more intense then those further up, so that it may be possible to read the sequence over only a small part of the gel. This usually indicates that there is an imbalance in the molar ratios of deoxy- and dideoxynucleotides, so that most chain-termination is taking place early on. It is a simple matter to run a series of reactions with varying ratios to find the optimum. Often this problem is restricted to one track (see *Figure 4*). A severe disturbance of the ratio can lead to the generation of artefact bands. Efficient sequencing of regions with markedly abnormal base composition may require compensating adjustments to the nucleotide mixes.

If an apparent imbalance in ratios appears, when previously there had been no problem, then it is likely that one of the components of the reaction mixes is deteriorating. Although

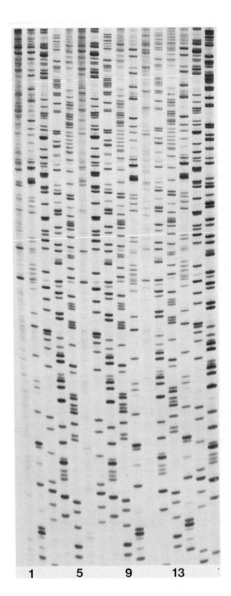

Figure 4. Effect of nucleotide mixes on sequencing reactions. **Lanes 1−4**; T,C,G,A tracks with twice usual amounts of dT in T nucleotide mix. **Lanes 5−8**; T,C,G,A tracks with twice usual amounts of dC in C nucleotide mix. **Lanes 9−12**; T,C,G,A tracks with twice usual amounts of dG in G nucleotide mix. **Lanes 13−16**; T,C,G,A tracks with usual amounts of nucleotides in all four mixes (for composition of nucleotide mixes see Chapter 2, *Table 2*).

the effect of this is likely to be greatest in one track, it can be expected to reduce the quality of other tracks too. When beginning sequencing for the first time it is wise to buy fresh stocks of nucleotides, aliquot the stock solutions and avoid freeze-thawing. Very often a nucleotide solution which has been found perfectly suitable for other biochemical purposes will be found to be unsuitable for sequencing purposes.

89

4.3.3 *Progressive overall reduction in band intensity with time*

Although the half-life of the ^{35}S nucleus is some three months, it should be borne in mind that the chemical stability of [α-^{35}S]dATP is notably less, and can be significantly reduced, as might be expected, by freeze-thawing repeatedly.

4.4 **Electrophoresis**

Problems encountered here usually arise as a result of incorrect preparation and loading of gel or samples, rather than in the gel electrophoresis itself. The resolution obtained with very thin gels (0.2 mm) is particularly sensitive to the nature of the sample, and the suppliers' protocols should be consulted carefully.

4.4.1 *Formation of bubbles while pouring the gel*

Pouring the thin gels used for sequencing requires some practice. To help avoid bubbles, ensure that the plates are very clean and grease-free, and run the gel solution between the two plates continuously. Try to avoid pausing, or interrupting the flow. Most X-ray film is thinner than most gel spacers, so if bubbles do form, it may be possible to dislodge them with a long strip of X-ray film inserted between the plates.

4.4.2 *Dark specks on the autoradiograph, often with thin lines extending downwards*

This is caused by dust on the gel plates prior to pouring the gel. Some types of paper towel used for drying plates leave a great deal of dust. See *Figure 5*.

4.4.3 *Disintegration of sample wells*

This is caused by a failure of polymerization and results in part of the well being washed away when the comb is removed and buffer added. It may be due to insufficient ammonium persulphate or TEMED, not leaving the gel for long enough to polymerise, or air getting to the wells. If the latter is the case, wrapping the top of the gel in Saran Wrap while polymerization is occurring may help. Degassing the gel mix may also help, as dissolved oxygen inhibits polymerization.

4.4.4 *Bands on gel very fuzzy, or even not discernible*

This is a common problem, and may have several causes. It can result from not leaving the gel for long enough for polymerization to be completed, and this should be suspected if the wells do not form properly. The time needed will depend on the exact amounts of ammonium persulphate and TEMED used as well as ambient temperature, but as a rough guide, the gel should be left for at least 30 min after pouring before use. Degassing the gel mix may help.

A frequent cause is the use of inadequate gel materials. The highest grades of reagents should be used (e.g. 'Electran' acrylamide), and filtering and deionizing carried out. Warming of the gel mix to dissolve the urea is not advised, and should certainly be very gentle indeed (no more than about 40°C). Excessive heating may reduce resolution later or even bring about spontaneous polymerization. Gel mixes should be stored at 4°C. Storage of mixes (especially those for buffer gradient gels) for more than a month may also lead to deterioration.

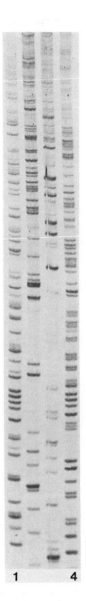

Figure 5. Effect of dirty gel plates on appearance of sequencing gel. **Lanes 1−4**; T,C,G,A tracks of template; note dark specks/streaks particularly visible here in the G track.

Gel running buffers should also be made up reasonably frequently (storage of 10 × TBE for more than a few days causes a precipitation). Some sequencers find that a deterioration of buffer quality causes a localized fuzziness, towards the top of the readable sequence, rather than throughout the gel. When carrying out prolonged gel runs, it is wise to circulate the buffer or change it half-way through the run.

Excessive heating of sample before loading can cause a combination of effects. Bands may be fuzzy, but they may also be faint, and the gel may also show a high level of

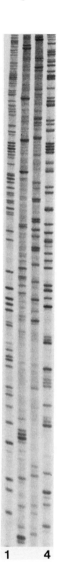

1   4

Figure 6. Effect of overheating sequence reactions before loading. **Lanes 1−4**; T,C,G,A tracks of reactions loaded after 10 min of incubation at 100°C.

artefact bands and a darkish background in the tracks (see *Figure 6*).

Before loading the samples into the wells, it is important to flush out any urea that has diffused out of the gel. Significant amounts of urea can diffuse out in quite a short time, so this should be done immediately prior to loading. It can be done by filling a Pasteur pipette or a syringe fitted with a Gilson tip with buffer and squirting the contents into the well. Failure to do this results in fuzzy and rather uneven bands (see *Figure 7*). Loading too much sample in the well will also cause fuzzy bands. As a general

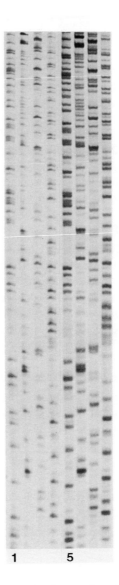

Figure 7. Effect of poor gel loading on appearance of sequencing gel. **Lanes 1−4**; T,C,G,A tracks showing effect of presence of excessive urea in gel wells. **Lanes 5−8**; T,C,G,A tracks showing effect of loading too large a volume of sample in gel wells prior to electrophoresis.

rule, this will happen if the well is loaded more than about half as deep as it is wide. Samples should be loaded as rapidly as possible and the power switched on immediately after loading is complete, to prevent reannealing of the DNA in the wells.

4.4.5 *'Smiling' in gels*

This is the name given to changes in mobility of oligonucleotides across the width of the gel, so that they run faster in the middle than at the edges. This means that bands which should be at the same level would form a 'U' across the gel. It is caused by variation in temperature across the gel, and can usually be avoided by clamping a metal sheet (2 mm thick aluminium is ideal) to the exposed gel plate, although some workers find this results in less sharp bands. Thermostatted plates are also available (at a price) for some sequencing gel apparatuses. Smiling can usually be allowed for in gel reading without much difficulty but may be a problem in automated sequencing.

4.4.6 *Difficulty in drying gels down for autoradiography*

This can be an indication that the urea has not been sufficiently leached from the gel during fixation for autoradiography. It may also render the bands diffuse. Fixing for 15 min is usually sufficient. Note that close contact between the gel and the film is needed during autoradiography, or the bands will become fuzzy and very faint.

5. SEQUENCE-DEPENDENT PROBLEMS

These are manifested (especially when using Klenow polymerase) as localized regions of sequence which are difficult to read easily. Certain rules can be applied and are set out in more detail in ref. 9 (and see Chapter 2, Section 3.3), but in general it is advisable to ensure that the same region is sequenced on the complementary strand, when the same problem will usually not arise. Sequencing on the complementary strand is of course necessary in any case.

5.1 **Variations in band intensity**

(i) *A bands*. In a run of As the bottom band is frequently the strongest.

(ii) *C bands*. Where two or more Cs are adjacent, the lowest is generally much weaker than the next one, so that the former may be scarcely visible. Individual C bands may also be very faint, less so if preceded by a G.

(iii) *G bands*. G may be weak in the sequence TG.

5.2 **Artefact bands**

As well as the non-specific artefacts already discussed, a few sequence-specific artefacts are sometimes (but not always) seen.

(i) *TGCC*. This sequence may cause an artefact band in the C track at or between the levels of the T and G bands.

(ii) *GCA*. Here there may be an artefactual T or C at the level of the A band.

5.3 **Compressions**

These are probably caused by G:C hairpins forming localized secondary structure in the DNA, persisting even under the conditions of electrophoresis. This secondary structure causes oligonucleotides to behave as though they were shorter than is actually the case, and thus migrate faster. It is diagnosed by bands running very close together, sometimes superimposed, usually with a gap or increased band spacing in the region above. There are three solutions to this problem. One is to make the conditions under which the gel is run more denaturing, which can be done by running at higher power (although this may also decrease resolution of the gel, and can cause plates to crack, particularly if they are chipped or scratched) or including formamide in the gel to a final concentration of 25−50%. The use of high formamide concentrations may lead to a higher background in the gel, however. Another solution is to sequence the complementary strand, when the position of the compression will usually have shifted a few bases. The third is to use ITP instead of GTP, as I:C base pairs are weaker than G:C ones, and secondary structures will therefore be less stable. For protocols for this, see Chapter 2, Section 3.2.1. Secondary structure may also lead to termination of synthesis by the polymerase, and may lead to compressions or pile-ups. Running the sequencing reactions at a higher temperature may alleviate this problem.

5.4. Pile-ups

These are also known as 'walls'. They are diagnosed as strong stops in all four tracks, often at two or more consecutive positions, and are generated during the sequence reactions rather than electrophoresis. Their occurrence may be dependent on the quality of template and enzyme, probably due to the presence of salt. Sequencing the complementary strand usually resolves the problem. (See also the previous section.) Carrying out the reactions at higher temperatures, using a thermostable polymerase ('Taq' polymerase, Cetus) may help (see Chapter 2, Section 3.4).

6. KEY

6.1 Plaque generation

Insufficient white plaques	Incorrect insert/vector ratio (2.1.1) Poor insert material (2.1.1) Incorrect phosphatasing (2.1.1, 2.1.2) Ligase/buffer faulty (2.1.3) Inefficient vector digestion (2.1.4) Nuclease contamination (2.1.1, 2.1.9)
Too many whites	Contamination of vector, buffers (2.1.6) No Xgal/IPTG or failure of colour (2.1.8) Contamination of host (2.2.2, 2.1.7)
Plaques too small	Lawn seeded too heavily (2.2.2) Insert too large/deleted (2.2.2)
Confluent plaques/very thin lawn	Wet plates (2.2.1) Contamination of host (2.2.2, 2.2.3) Cells overheated (2.2.2)
Mottled lawn	Agar too cool (2.2.1)
No plaques	Cells not competent (2.2.2) Pili lost/strain contaminated (2.2.2) Nuclease contamination (2.1.9)

6.2 **Sequencing reactions, gels**

Artefact bands in all tracks	Dirty template (4.1.3)
	Poor quality polymerase (4.3.1)
	Excessive freeze-thawing (4.1.3, 4.2.1)
	Cells grown too long/too heavily seeded (3, 4.1.3)
	Excess primer (4.2.2)
Artefact bands in one track	Imbalanced nucleotide mix (4.3.2)
	DTT old (especially C track) (4.3.1)
Duplicate bands	Mixed template (4.1.3)
	Cross-contamination of mixes (4.3.1)
Bands faint/absent	Deletion in phage (4.1.2)
	Low yield of DNA (3, 4.1.1, 4.1.2)
	Poor nucleotide mixes/radioisotope (4.3.2, 4.3.3)
	Faulty primer (4.2.1)
	Freeze-thawing of template (4.1.3)
Bands fuzzy	Urea not washed out of wells (4.4.4)
	Too much sample loaded (4.4.4)
	Poor quality/stale electrophoresis reagents (4.4.4)
	Incomplete polymerization of gel (4.4.3)
	Gel not processed correctly before autoradiography (4.4.6)
Background dark, bands fuzzy	Samples overheated (4.4.4)
	Plates dirty (4.4.2)
Gel difficult to read, though bands distinct	Smiling (4.4.5)
	Compression (5.3)
	Pile-up (5.4)

7. ACKNOWLEDGEMENTS

We are grateful to Drs Duncan Moore and Martin Maiden for their helpful suggestions and advice.

8. REFERENCES

1. Maniatis,T., Fritsch,E.F. and Sambrook,J. (1982) *Molecular Cloning.* Cold Spring Harbor Laboratory, Cold Spring Harbor, New York.

2. Davis,L.G., Dibner,M.D. and Battey,J.F. (1986) *Basic Methods in Molecular Biology*. Elsevier, New York.
3. Ausubel,F.M., Brent,R., Kingston,R.E., Moore,D.D., Seidman,J.G., Smith,J.A. and Struhl,K. (1987) *Current Protocols in Molecular Biology*. Wiley-Interscience, New York.
4. Dretzen,G., Bellard,M., Sassone-Corsi,P. and Chambon,P. (1981) *Anal. Biochem.*, **112**, 295.
5. Adams,D.S., Lurhman,R. and Lizardi,P.M. (1985) *Gene Anal. Techn.*, **1**, 109.
6. Girvitz,S.C., Bacchetti,S., Rainbow,A.J. and Graham,F.L. (1980) *Anal. Biochem.*, **106**, 492.
7. Messing,J. (1979) *NIH Recombinant DNA Technical Bulletin*, **2**, 43.
8. Gibson,T. (1984) PhD thesis, University of Cambridge.
9. Bankier,A.T. and Barrell,B.G. (1983) In *Techiques in the Life Sciences: Nucleic Acid Biochemistry*. Elsevier Scientific Publishers, Ireland Ltd, p. 1.

CHAPTER 4

Sequencing of double-stranded DNA

G.MURPHY and E.S.WARD

1. INTRODUCTION

Since the introduction of the chain-termination method of DNA sequencing (1) the vast majority of applications of this method have probably involved single-stranded circular DNA from M13 bacteriophages (2,3,4). However, it has long been appreciated that there are considerable advantages in direct sequencing of double-stranded DNA. Long inserts, particularly of repetitive sequences, are often unstable in M13 phages but are usually stable in plasmid vectors. The sequencing of inserts in plasmids avoids the need for tedious subcloning into M13 phage and can be used to test the fidelity of constructions directly. Short inserts can often be sequenced completely by priming from both ends, avoiding the need for recloning into vectors with the polylinker in the opposite orientation or synthesis of the complementary strand of single-stranded inserts followed by repriming with primer to the complementary strand and chain-termination sequencing (5). Double-stranded DNA is also amenable to a number of directed deletion sequencing methods and the template products of such strategies can be easily and accurately screened for size, allowing a nested set of deletions to be selected.

For these reasons many attempts have been made to apply enzymatic chain-termination methods to double-stranded DNA (6−8). The number of bases of a plasmid insert which can now be sequenced by these methods is becoming comparable to that for single-stranded phages, although the overall sequencing quality, and hence accuracy, is still lower. However, because of the advantages outlined above, sequencing of plasmid DNA has become a powerful technique likely to be used increasingly as methods are improved.

DNA sequencing can also be applied to double-stranded DNA which has not been 'cloned' in the traditional sense, but obtained directly from the organisms under study, often using amplification by the 'polymerase chain reaction' (PCR). This is described in Section 3.

2. SEQUENCING OF DOUBLE-STRANDED CLONED DNA

The quality of the sequence obtained from double-stranded DNA, both in the number of bases and accuracy, depends to a very large extent on the purity of the DNA and rigorous application of the methods described below. There is often very little flexibility either in the quantity of reagents which can be added or in the timing of reactions. Where several different methods can be applied to a particular step the variations are given, but the convention used is that the first method is known to be the most efficient or productive. The other variations are included because of the natural conservatism of molecular biologists in such methods as plasmid isolation.

Table 1. Preparation of plasmid from minipreps by boiled lysis.

1.	Toothpick a colony into 10 ml of LB[a] in a flat-bottomed, screw-top bottle and grow for $15-18$ h at 37°C, shaking at $200-250$ r.p.m. and using antibiotic selection if required.
2.	Centrifuge the cells at 1500 g for 10 min. Decant the supernatant and leave the inverted bottle to drain for 5 min on absorbent paper.
3.	Resuspend the cells by vortexing vigorously in 100 μl of 50 mM Tris$-$HCl pH 8, 25% sucrose, then transfer the suspension to a 1.5-ml microcentrifuge tube.
4.	Add 600 μl of MSTET[b] solution, then spot 14 μl of 40 mg ml^{-1} lysozyme in 50 mM Tris$-$HCl pH 8, 50% (v/v) glycerol[c] onto the inside of the tube.
5.	Cap the tubes, mix the contents thoroughly by shaking and transfer immediately to a boiling water bath for 1 min.
6.	Place the tubes on ice for 1 min, then centrifuge at 10 000 g for 30 min at 4°C.
7.	Remove the gelatinous pellet with a toothpick. Alternatively, to reduce contamination with chromosomal DNA, use a micropipette to transfer the supernatant to a fresh tube.
8.	Add 60 μl of 3 M sodium acetate, pH 5, and 600 μl of isopropanol and leave on ice for 5 min before centrifuging at 10 000 g for 10 min. Aspirate off the supernatant, recentrifuge for 5 sec and remove all traces of liquid. Resuspend in 200 μl of TE[d].
9.	Extract the sample with 100 μl each of phenol and chloroform:isoamyl alcohol (25:1;v/v), vortexing vigorously for 1 min followed by centrifugation at 10 000 g for 4 min. Remove the supernatant, avoiding any contamination with the white interface, and repeat the phenol/chloroform extraction.
10.	Add 20 μl of 3 M sodium acetate, pH 5, and 550 μl of ethanol. Leave the tubes for 1 h at -70°C or at -20°C overnight. Centrifuge for 10 min at 10 000 g, then remove the supernatant and rinse the pellet with 500 μl of 80% ethanol stored at -20°C. Centrifuge for 2 min, aspirate off all traces of liquid and dry the pellet for 5 min under vacuum before dissolving the pellet in 50 μl of TE.

[a]LB is 1% Bactotryptone, 0.5% Bacto yeast extract and 1% (w/v) NaCl.
[b]MSTET is 50 mM Tris$-$HCl pH 8, 50 mM Na$_2$EDTA, 5% (v/v) Triton X-100 and 5% (w/v) sucrose.
[c]The lysozyme solution is stable at -20°C for at least 2 weeks.
[d]TE is 10 mM Tris$-$HCl pH 8, 1 mM Na$_2$EDTA.

In the 'shotgun' approach to single-stranded DNA sequencing relatively short lengths of DNA are sequenced randomly, so that in a large sequencing project the same stretch of DNA may be sequenced up to six times. More recently methods have become available to increase the length of the oligonucleotide chains formed in a sequencing reaction to 1000 or more bases. The use of these methods provides two approaches to sequencing double-stranded DNA:

(i) to aim to generate readable sequence of $300-350$ bases at a time, and when sequencing large inserts to generate a size-nested set of deletions by directed deletion techniques,

(ii) to attempt to obtain as much sequence information as possible from any insert, by synthesizing long oligomer chains and separating them on several gels of different acrylamide concentrations and with extended run times.

With either approach some additional effort has to be invested, either in generating and screening deletions or in handling long gels. Deciding which approach to use is largely subjective. However, using deletion techniques and reading $300-350$ bases per template does provide security, in that as in 'shotgun' techniques the sequence may be confirmed by sequencing the same region several times. The methods initially described here favour the first approach, but alternative protocols are given for those tending towards the second.

Table 2. Preparation of plasmid from minipreps by alkaline lysis.

A. Modified Birnboim and Doly (10) procedure.

1. Grow and centrifuge cells as described in *Table 1*.
2. Suspend the cells in 200 μl of 25 mM Tris−HCl pH 8, 10 mM Na$_2$EDTA, 50 mM glucose and 4 mg ml^{-1} lysozyme. Incubate for 5 min at room temperature then mix gently.
3. Add 400 μl of 0.2 M NaOH, 1% (w/v) SDS and mix by inverting the tube several times before leaving on ice for 5 min.
4. Add 300 μl of 3 M potassium acetate, 2 M acetic acid and vortex gently to mix. Leave on ice for 5 min then centrifuge at 10 000 *g* for 10 min.
5. Remove 800 μl of the supernatant, being careful not to disturb the pellet, and transfer to a fresh tube. Add 200 μl each of phenol and chloroform: isoamyl alcohol[a], vortex for 1 min and centrifuge for 4 min at 10 000 *g*.
6. Transfer the supernatant to a fresh tube containing 800 μl of isopropanol. Mix by vortexing and leave on ice for 5 min before centrifugation at 10 000 *g* for 10 min. Aspirate off the supernatant, centrifuge for 2 sec and remove all liquid with a drawn-out pasteur pipette. Dry under vacuum for 10 min.
7. Resuspend the pellet in 100 μl of TE, add 3 μl of 10 mg ml^{-1} RNase A[b] and incubate for 20 min at 37°C.
8. Add 50 μl each of phenol and chloroform[a], then vortex and centrifuge as above. Transfer the supernatant to a fresh tube containing 10 μl of sodium acetate pH 5 and add 300 μl ethanol. Continue as in *Table 1*.

B. Rapid isolation method.

1. Grow the cells and centrifuge them as in (A). Resuspend each pellet in 250 μl of 25 mM Tris−HCl pH 8, 25 mM Na$_2$EDTA, 0.3 M sucrose and 2 mg ml^{-1} lysozyme and incubate on ice for 30 min.
2. Add 250 μl of 0.3 M NaOH, 2% SDS with immediate vortexing. Incubate for 15 min at 70°C then cool to room temperature in a water bath.
3. Add 80 μl of unbuffered phenol/chloroform[c], vortex for 30 sec and centrifuge at 10 000 *g* for 4 min.
4. Carefully transfer the upper phase to a fresh tube containing 70 μl of unbuffered 3 M sodium acetate and 700 μl of isopropanol. Vortex to mix and leave on ice for 5 min followed by centrifugation at 10 000 *g* for 10 min. Aspirate off the supernatant, recentrifuge briefly and remove all the liquid by aspiration.
5. Dissolve the pellet in 100 μl of TE and vortex for 1 min with 50 μl each of phenol and chloroform[a]. Centrifuge for 4 min and transfer the upper phase to 10 μl of unbuffered 3 M sodium acetate and add 110 μl of isopropanol. Leave on ice and centrifuge as above before rinsing the pellet in 500 μl of cold ethanol, drying and dissolving in 50 μl of TE.

[a]See *Table 1* stage 9 for preparation.
[b]The RNase is dissolved in 25 mM Tris−HCl pH 8 and 10 mM NaCl. Boil for 10 min and allow to cool slowly to room temperature. Store at −20°C.
[c]Dissolve 5 g of phenol in 5 ml of chloroform, 1 ml of H$_2$O and 5 mg of 8-hydroxyquinoline.

From this point the sequencing of double-stranded DNA will be exemplified by the use of plasmid DNA, but it should be recognized that the same techniques can be applied to the replicative form of M13 phage or lambda phage. Protocols for generating deletion sets and ? cloning strategies are given in Chapter 1.

2.1 Plasmid isolation

The two most popular methods of plasmid isolation are the boiled-lysis method of Holmes and Quigley (9) and the alkaline-lysis method of Birnboim and Doly (10). The use of these methods for the preparation of plasmid from minipreps is described in *Tables*

Table 3. Large-scale plasmid isolation.

1.	Grow cells overnight in 400 ml of LB[a] with antibiotic selection if required. Centrifuge at 1500 g for 10 min, decant off the supernatant and invert container to drain for 2 min on an absorbent pad.
2.	Resuspend the cells in 1 ml of 25 mM Tris−HCl pH 8, 10 mM Na_2EDTA and 50 mM glucose. Transfer to a 50-ml centrifuge tube and add 7 ml of the same solution containing 2 mg ml^{-1} lysozyme. Leave on ice for 30 min with occasional gentle mixing.
3.	Add 16 ml of 0.2 M NaOH, 1% (w/v) SDS, seal the tube and mix by inverting several times before leaving for 10 min on ice.
4.	Add 12 ml of 3 M potassium acetate, 2 M acetic acid, seal the tube and shake vigorously to mix. Leave on ice for 30 min then centrifuge at 10 000 g for 10 min.
5.	Decant supernatant into a 100-ml centrifuge tube through Miracloth[b] to remove any floating particles. Add ethanol to fill the tube, seal and mix by inverting. Leave at −20°C for 30 min.
6.	Collect the flocculent precipitate by centrifugation at 10 000 g for 5 min. Decant off the supernatant and drain the tube by leaving inverted for 5 min on an absorbent pad. Dissolve the pellet in 10 ml of TE, transfer to a 50-ml centrifuge tube on ice and add 5 ml of 7.5 M ammonium acetate. Leave for 30 min and then centrifuge at 10 000 g for 10 min.
7.	Pour off the supernatant into a fresh 50-ml centrifuge tube and add 150 μl of 1 M $MgCl_2$ and 30 ml of ethanol, mix and leave at −20°C for 30 min. Centrifuge for 10 min at 10 000 g, remove the supernatant and rinse the pellet with 5 ml of cold 80% ethanol. Dry under vacuum for 5 min.
8.	Add 2 ml of 10 × TE and dissolve the pellet before adding 3.22 g of CsCl and 0.6 ml of 10 mg ml^{-1} ethidium bromide. Make up the volume to 4.3 ml before centrifuging at 12 000 g for 20 min. Transfer the supernatant to an appropriate centrifuge tube, fill the tube with rebanding solution[c] and centrifuge at 200 000 g for 15 h.
9.	Visualize the plasmid band in long wave UV and remove by inserting a 21G needle on a 1-ml syringe into the tube below the plasmid band. Transfer ∼ 1 ml of solution to a fresh centrifuge tube, make up the volume with rebanding solution and centrifuge for 6 h at 20 000 g. Remove the plasmid band as above, and transfer to a 15-ml siliconized centrifuge tube.
10.	Remove the ethidium bromide by extracting 3−4 times with 3 ml of isoamyl alcohol saturated with 50 mM Tris−HCl pH 8, 1 mM Na_2EDTA, vortexing to mix the phases. Add 3 vol of H_2O to the sample, followed by 40 μl of 1 M $MgCl_2$ and 8 ml of ethanol. Leave at −20°C for 1 h and then centrifuge at 10 000 g for 10 min.
11.	Dissolve the pellet in 0.4 ml of TE and add 40 μl of 3 M sodium acetate, 5 μl of 1 M $MgCl_2$ and 1.1 ml of ethanol. Leave at −20°C for 1 h, centrifuge at 10 000 g for 10 min. Rinse the pellet with 2 ml of cold ethanol, dry under vacuum for 5 min and dissolve in TE.

[a]See *Table 1*.
[b]Calbiochem.
[c]Dissolve 76.62 g of CsCl and 9.53 ml of 10 mg ml^{-1} ethidium bromide in 10 × TE to a volume of 100 ml.

1 and *2*, which include a rapid alternative method (11), while a large-scale alkaline-lysis method employing caesium chloride gradient purification of plasmid DNA is outlined in *Table 3*. All of these methods produce sequencing-quality DNA, but the boiled-lysis method appears to produce templates which generate fewer artefacts upon sequencing. Caesium chloride gradient purified DNA provides a useful control against which other techniques can be judged, as the DNA is far less contaminated (e.g. with chromosomal DNA, RNA) than miniprep DNA. However, some background is often observed when sequencing gradient purified DNA, perhaps due to some nicking of DNA in ethidium bromide solutions when exposed to UV light.

The amount of DNA obtained from minipreps is variable and depends on plasmid copy number, but as a rough guide one would expect 25−30 μg from a 10-ml boiled-lysis preparation and about 20−25 μg using alkaline lysis.

The preparation of RF-DNA from M13 phage-transfected cells is described in *Table*

Table 4. Preparation of M13 RF-DNA.

1.	Grow a culture of a suitable host in 10 ml of minimal medium[a] at 37°C with shaking overnight.
2.	Transfer 1 ml of the culture to 10 ml of YT[b] and shake at 37°C for 1 h before adding a single plaque from M13 transfected cells. Grow for 6 h and then centrifuge the cells at 1500 *g* for 10 min.
3.	The cells are then treated as for a plasmid miniprep (*Table 1*). If a large-scale RF prep is required, inoculate 400 ml of YT with 4 ml of the overnight cell culture in minimal medium and grow until the cells reach an OD_{640} of $0.5-0.6$. Add the supernatant from step 2 to the culture and grow for a further 4 h. Centrifuge the cells at 1500 *g* for 10 min and continue as in *Table 3*. The yield should be $200-400$ μg per litre of culture.

[a]Minimal medium ingredients are autoclaved separately then mixed aseptically. Mix together 887 ml of water, 10 ml of 20% (w/v) glucose, 1 ml of 1 M $MgSO_4$, 1 ml of 0.1 M $CaCl_2$, 1 ml of 1 M thiamine$-$HCl and 100 ml of M9 salts (7 g of Na_2HPO_4, 3 g of KH_2PO_4, 1 g of NH_4Cl and 0.5 g of NaCl in a total volume of 100 ml).
[b]YT is 8 g of Bactotryptone, 5 g of yeast extract and 5 g of NaCl in a volume of 1 litre.

Table 5. Chain-termination mixes for sequencing.

1.   Mix the following amounts of 0.5 mM dNTPs and 5 mM ddNTPs (dissolved in T0.1E buffer[a]):
If using 7-deaza-dGTP, substitute for dGTP in same molar quantities.

	A mix	*C mix*	*G mix*	*T mix*
dCTP	250 μl	12.5 μl	250 μl	250 μl
dGTP	250 μl	250 μl	12.5 μl	250 μl
dTTP	250 μl	250 μl	250 μl	12.5 μl
ddATP	1.5 μl (3 μl for [^{32}P]dATP)		$-$	$-$
ddCTP	$-$	8 μl	$-$	$-$
ddGTP	$-$	$-$	16 μl	$-$
ddTTP	$-$	$-$	$-$	50 μl
T0.1E[a]	250 μl	480 μl	470 μl	440 μl

[a]T0.1E is 10 mM Tris$-$HCl pH 8, 0.1 mM Na_2EDTA.

4. The yields of RF are lower than from plasmid minipreps because of the low copy number per cell; typically 4 μg per 10 ml of culture.

2.2 Sequencing

2.2.1 *Equipment*

Sequence reactions are performed with up to 10 templates in a set. Reagents are dispensed onto the walls of 1.5-ml lidless centrifuge tubes held in Eppendorf 10 place racks and mixed by brief centrifugation in an Eppendorf 5413 centrifuge. The reagent concentrations are adjusted so that multiples of 2 μl can be dispensed rapidly using a Hamilton PB600 repeating dispenser.

2.2.2 *Standard chain-termination mixes*

The mixes of dNTPs and ddNTPs used in the standard sequencing protocol are given in *Table 5*. With these mixes good sequence can be obtained from a few bases from the end of the primer up to 350 bases from it. Mixtures which can be used in deviations from the standard protocol for sequencing either very close to the end of the primer or $700-800$ bases from it are described in Sections 2.2.6 and 2.2.7.

Table 6. Preparing templates for sequencing.

A. Spin-dialysis method

1. Mix 18 μl (12−15 μg of DNA) of a plasmid miniprep with 2 μl of RNase A (*Table 2*) and incubate for 20 min at 37°C. Add 5μl of 1 M NaOH, 1 mM Na$_2$EDTA and incubate for 15 min at 37°C. If using caesium chloride gradient purified DNA use 15 μg in 20 μl of TE without RNase treatment.
2. Add the sample to the top of the gel of a prepared spin-dialysis tube (*Table 7*), being careful not to disturb the gel layer. Centrifuge at 200 *g* for 4 min and use the dialysate immediately.

B. Precipitation method

1. Follow the steps in (A) above, but instead of applying the sample to a spin-dialysis tube, neutralize it by the addition of 2.5 μl of 2 M ammonium acetate pH 4.5.
2. Add 100 μl of ethanol stored at −20°C and leave the sample at −70°C for 10 min. Centrifuge at 10 000 *g* for 10 min, remove the supernatant and rinse the pellet with cold ethanol.
3. Dry the pellet under vacuum for 5 min and dissolve in 25 μl of T0.1E[a] and use immediately.

C. Linearization method

1. Mix 18.5 μl of plasmid miniprep DNA with 2 μl of RNase A (*Table 2*), or use 15 μg of caesium chloride gradient purified plasmid DNA in 20.5 μl of TE, together with 2.5 μl of 10 × restriction buffer and 2 μl (20 U) of a restriction enzyme cutting the polylinker region on the opposite side of the insert to the primer site. Incubate at 37°C for 30 min.
2. Boil the sample for 5 min, cool rapidly on ice and use immediately.

[a]10 mM Tris−HCl pH 8, 0.1 mM Na$_2$EDTA.

Table 7. Preparation and testing of spin-dialysis tubes.

A. Preparation

1. Equilibrate Sepharose-CL6B[a] in T0.1E[b] and adjust the buffer volume to produce a packed gel:buffer supernatant ratio of 2:1.
2. Pierce the base of a 0.5-ml centrifuge tube with a 21G needle so that about 2/3 of the needle bevel emerges. Place the tube inside a 1.5-ml centrifuge tube completely pierced through the bottom with the same needle.
3. Add 25 μl of a slurry of 200 micron glass beads[c] in water to the 0.5-ml tube, followed by 300 μl of the Sepharose slurry.
4. Place the assembly into a 9-mm internal diameter tube and centrifuge at 200 *g* for 4 min. Transfer the 0.5-ml tube to an intact 1.5-ml test-tube and use within an hour to prevent drying of the gel matrix.

B. Testing

1. Prepare tubes as in A.
2. Add 25 μl of 10 mg ml^{-1} Blue Dextran 2000[a] and 10 mg ml^{-1} Orange G dye in TE to the top of the Sepharose and centrifuge as above.
3. Transfer the spin-column to a fresh tube, add 25 μl of TE and recentrifuge. More than 90% of the Blue Dextran should be present in the first dialysate, while no Orange G should pass through in the second.

[a]Pharmacia.
[b]10 mM Tris−HCl pH 8, 0.1 mM Na$_2$EDTA.
[c]Jencons Ballotini beads, No. 11.

2.2.3 *Denaturation of template DNA*

In order for the primer to be able to bind to the priming site the two DNA strands must first be separated. The two main methods are denaturation by alkali (8) and linearization of the plasmid by cutting with a restriction enzyme followed by denaturation

Table 8. Sequencing reactions.

1.	If sequencing with $[\alpha\text{-}^{35}S]dATP$, 8.5 μl of the prepared template is added to 1 μl of 10 $\times$ TM[a] and 1 μl of 10 μg ml^{-1} primer. For sequencing with $[\alpha\text{-}^{32}P]dATP$ use 5 μl of template with the same amount of primer and 10 $\times$ TM and make the final volume up to 10.5 μl with water.
2.	Incubate at 37°C for 15 min and centrifuge the tubes briefly to spin down any condensation.
3.	Dispense 2.4 μl to each of four tubes, followed by 2 μl of chain-termination mix.
4.	For 10 templates mix 9 μl of 10 $\times$ TM with 1 μl of 100 mM DTT, 66 μl of H$_2$O, 4 μl (20 U) of Klenow fragment of DNA polymerase 1[b] and 10 μl of $[^{35}S]$- or $[^{32}P]dATP$[c]. Mix thoroughly and quickly dispense 2 μl of the mixture onto the wall of each tube. Centrifuge to mix, place immediately in a water bath at 42°C and incubate for 10 min.
5.	Spot 2 μl of a chase mix (0.5 mM in each of dATP, dCTP, dGTP and dTTP) onto the wall of the tubes and centrifuge to mix. Return the samples to the bath and continue the incubation for 5 min.
6.	Terminate the reaction by the addition of 4 μl of formamide dye mix, consisting of 10 ml of deionized formamide, 200 μl of 0.5 M Na$_2$EDTA pH 8 and 10 mg each of xylene cyanol and bromophenol blue.
7.	Immediately before loading onto an acrylamide gel denature the samples by boiling for no more than 2 min, then load 1.5−2 μl of each sample onto the gel.

[a]10 $\times$ TM is 100 mM Tris−HCl pH 8, 50 mM MgCl$_2$.
[b]Boehringer Mannheim 5 Uμl^{-1}.
[c]$[\alpha\text{-}^{35}S]dATP$ at 500 Ci mmol^{-1}, $[\alpha\text{-}^{32}P]dATP$ at 800 Ci mmol^{-1}, both at 10 mCi ml^{-1}.

by boiling (6,7), both described in *Table 6*. Conventionally, use of the alkali denaturation method involves neutralization of the sample followed by ethanol precipitation and recovery of the DNA by centrifugation. The time-consuming processes of restriction enzyme digestion or ethanol precipitation may be avoided by neutralizing the alkali-treated DNA and recovering the sample in its original volume by passage through a spin-dialysis column, the preparation of which is described in *Table 7*. The use of spin-dialysis confers the additional advantage of cleaning up the template by removing traces of low molecular weight compounds which may interfere with the sequencing reactions. This method is preferable to the two other methods in that:

(i)   it is rapid,

(ii) precipitation of the DNA can cause sequencing artefacts if the DNA becomes contaminated with salt, and

(iii) DNA linearized by digestion often produces strong banding across all four lanes, the position of the artefact changing as the restriction enzyme is altered.

The amount of DNA used in the denaturation provides enough material to sequence a short insert from both ends, leaving sufficient DNA for a further reaction if required. Denatured templates can be stored frozen at −20°C for at least several weeks. The amount of starting material can be adjusted easily within the precipitation or restriction enzyme protocols, but because of possible problems of low recovery of DNA it is unwise to use spin-dialysis for samples smaller than a final volume of 20 μl i.e. using less than 80% of the volumes given in *Table 6*.

2.2.4 *Chain-termination reaction*

Following denaturation the template is annealed to the primer by brief incubation at 37°C (*Table 8*). There is no advantage in annealing at higher temperatures and slow cooling to room temperature, since this may cause problems through reannealing of the complementary strands and premature termination. The amount of primer indicated

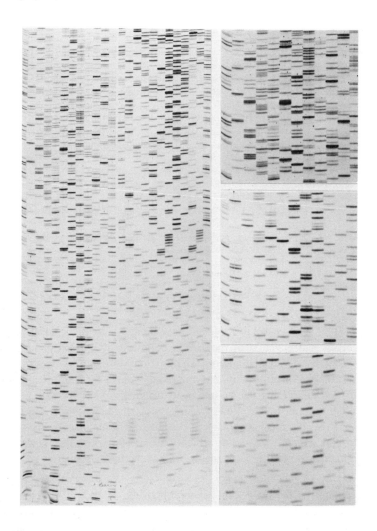

Figure 1. Supercoiled DNA sequencing gel. Three templates were loaded in the order A-C-G-T onto a 6% acrylamide gel and electrophoresed for 2 h before a second loading and electrophoresing for a further 90 min. The **left panel** is the entire gel with the first set of samples in the first 12 tracks. The **bottom right panel** shows the region centred on 50 bases from the gel bottom, the **middle right panel** the region centred on 150 bases and the **top right panel** the region centred on 250 bases.

is equimolar for 5 μl of a plasmid of 4 kb. Do not use larger amounts of primer, because at higher concentrations priming to sequences of lower specificity may occur, generating ghost bands on the gel (see Chapter 3, Section 4.2).

The reactions are performed at 42°C to reduce secondary structure formation in the template, particularly in regions of high G−C content. If such artefacts are observed the reaction temperature may be increased to 50°C, but if higher temperatures are used additional Klenow fragment of DNA polymerase I should be included in the chase-mix.

Table 9. High ddNTP/dNTP ratio chain-termination mixes.

1.　Prepare stock solutions by mixing:

	A0	*C0*	*G0*	*T0*
1 mM dCTP	100 μl	5 μl	100 μl	100 μl
1 mM dGTP	100 μl	100 μl	5 μl	100 μl
1 mM dTTP	100 μl	100 μl	100 μl	5 μl
T0.1E buffer	–	100 μl	100 μl	100 μl

Followed by:

(i)　4 × Stock
　　A mix: 37.5 μl A0 + 2 μl 1 mM ddATP + 60 μl T0.1E
　　C mix: 37.5 μl C0 + 3.2 μl 5 mM ddCTP + 60 μl T0.1E
　　G mix: 37.5 μl G0 + 6.4 μl 5 mM ddGTP + 56 μl T0.1E
　　T mix: 37.5 μl T0 + 20 μl 5 mM ddTTP + 40 μl T0.1E

(ii)　8 × Stock
　　A mix: 37.5 μl A0 + 4 μl 1 mM ddATP + 60 μl T0.1E
　　C mix: 37.5 μl C0 + 6.4 μl 5 mM ddCTP + 55 μl T0.1E
　　G mix: 37.5 μl G0 + 12.8 μl 5 mM ddGTP + 50 μl T0.1E
　　T mix: 37.5 μl T0 + 40 μl 5 mM ddTTP + 20 μl T0.1E

2.　Perform the sequencing reactions as in *Table 8*, but substituting one of the above mixes for the standard chain-termination mix, and omitting bromophenol blue from the loading buffer.

3.　Separate the samples on a 20 × 50 cm gel using an 8% w/v acrylamide mix and wedge-shaped gel spacers varying from 0.4 mm at the top to 1.2 mm at the bottom. Use loading buffer with bromophenol blue in adjacent lanes to act as markers.

4.　Run the gel until the bromophenol blue is 5 cm from the bottom. Increase the fixing and drying times to 60 min each.

2.2.5 *Gel electrophoresis*

With the methods described in Section 2 and electrophoresis conditions as in Chapter 2, good, readable and unambiguous sequence extending to about 325 bases from the primer should be observed on a 50-cm, 6% acrylamide buffer gradient gel (12). However, strong bands are observed in regions above which the bands are not well separated (*Figure 1*). The number of readable bases can be increased by:

(i)　separating part of the sample on a 50-cm 6% acrylamide non-gradient gel for about 100 min,

(ii)　separating another part of the sample on a similar gel or a 5% acrylamide gel for about 4 h.

Using these methods the amount of readable sequence can be extended to around 450−500 bases. When extended separations are performed a region of blurred bands is frequently observed at around 350−400 bp. This is caused by a reaction between borate ions in the buffer and glycerol from the Klenow storage buffer and can be eliminated by passing the diluted enzyme mixture through a spin-dialysis column as described in *Table 7*, using Sepharose equilibrated in the reaction buffer, before the addition of [^{35}S]dATP.

Table 10. Sequencing in the kilobase range with the Klenow enzyme.

1.	Prepare a primed template and dispense 2.4 μl into each of four tubes as described in Sections 2.2.3 and 2.2.4.		
2.	Prepare an extension and labelling mix 7.5 μM in dCTP, dGTP and dTTP (EL mix). Mix for the following approximate range of bases synthesized:		

	10−500 bases	80−1000 bases	300−1200 bases
1 × TM buffer[a]	68 μl	65 μl	65 μl
0.1 M DTT	11 μl	11 μl	11 μl
EL mix	2.2 μl	4.5 μl	9 μl
Klenow enzyme (5 U μl^{-1})	4 μl	4 μl	4 μl
[^{35}S]dATP[b]	5 μl	5 μl	5 μl

3.	Dispense 2 μl of the appropriate mix into each reaction tube. Mix by centrifugation and incubate at 42°C in a water bath for 5 min.
4.	Add 2 μl of the appropriate termination mix per tube, centrifuge and return to the water bath for 5 min. Termination mixes:

	A mix	*C mix*	*G mix*	*T mix*
5 mM dATP	7 μl	70 μl	70 μl	70 μl
5 mM dCTP	70 μl	7 μl	70 μl	70 μl
5 mM dGTP	70 μl	70 μl	7 μl	70 μl
5 mM dTTP	70 μl	70 μl	70 μl	7 μl
5 mM ddATP	84 μl	−	−	−
5 mM ddCTP	−	28 μl	−	−
5 mM ddGTP	−	−	42 μl	−
5 mM ddTTP	−	−	−	140 μl
T0.1E	700 μl	750 μl	740 μl	640 μl

5.	Add 4 μl of formamide dye (*Table 8*) and mix by centrifugation. Denature by boiling for 2 min and load 2 μl onto a suitable gel.

[a]*Table 8*.
[b]500 Ci mmol^{-1}, 10 mCi ml^{-1}.

Table 11. Sequencing in the kilobase range with Sequenase.

1.	Proceed as in *Table 10*, with the exception that:
(i)	the volume of enzyme added is reduced to 3 μl (36 U Sequenase);
(ii)	the initial labelling reaction should be at room temperature;
(iii)	the size ranges obtained with the three labelling mix dilutions are around 10−250, 20−450 and 100−1000 bases respectively.
2.	The termination mixes are:

	A mix	*C mix*	*G mix*	*T mix*
5 mM dATP	25 μl	25 μl	25 μl	25 μl
5 mM dCTP	25 μl	25 μl	25 μl	25 μl
5 mM dGTP	25 μl	25 μl	25 μl	25 μl
5 mM dTTP	25 μl	25 μl	25 μl	25 μl
5 mM ddATP	2.5 μl	−	−	−
5 mM ddCTP	−	2.5 μl	−	−
5 mM ddGTP	−	−	2.5 μl	−
5 mM ddTTP	−	−	−	2.5. μl
T0.1E	900 μl	900 μl	900 μl	900 μl

Table 12. Resolving sequencing artefacts caused by secondary structure formation.

A. Use of dITP and Sequenase

1. Substitute 15 μM dITP in the extension and labelling mix of *Table 10*, in place of the dGTP.
2. Replace termination mix with that below.

	A mix	C mix	G mix	T mix
5 mM dATP	25 μl	25 μl	25 μl	25 μl
5 mM dCTP	25 μl	25 μl	25 μl	25 μl
5 mM dITP	50 μl	50 μl	50 μl	50 μl
5 mM dTTP	25 μl	25 μl	25 μl	25 μl
5 mM ddATP	2.5 μl	–	–	–
5 mM ddCTP	–	2.5 μl	–	–
0.5 mM ddGTP	–	–	5 μl	–
5 mM ddTTP	–	–	–	2.5 μl
T0.1E	875 μl	875 μl	875 μl	875 μl

3. Start the labelling reaction and immediately begin to dispense aliquots of the termination mixes onto the wall of the tubes. The labelling reaction should be carried out for as short a time as possible – ideally no more than 2 min.
4. Spin to mix and transfer immediately to a 37°C bath, then incubate for 3 min before the addition of formamide dye.

B. Stabilizing template with single-stranded DNA-binding protein

1. After denaturation of the template, spin-dialysis and priming, add between 1 and 5 μg (depending on the severity of the problem) of T4 Gene 32 protein[a] to the sample.
2. Following the sequencing reactions and the addition of formamide dye, add 0.5 μg of proteinase K[b] and incubate the mixture at 65°C for 20 min before samples are loaded onto the gel.

[a]Pharmacia.
[b]Boehringer.

2.2.6 *Sequencing close to the primer site*

When sequence adjacent to the primer is required, for example when it is necessary to sequence through the junction of vector and insert DNA to check that the sequence is in the correct protein coding frame, some modifications to the standard protocol are necessary (*Table 9*). By increasing the ddNTP/dNTP ratio chain termination occurs more easily, generating strong bands on the autoradiograph next to the primer site. To sharpen the bands the acrylamide concentration is increased to 8% and a wedge-shaped gel, 0.4 mm at the top and 1.2 mm at the bottom, is employed. Bromophenol blue should not be included in the formamide dye, as it has a deleterious effect on the separation and resolution of short oligomers. Instead a small amount of dye containing bromophenol blue should be loaded in spare tracks on the outer edges of the gel.

2.2.7 *Kilobase sequencing*

In conventional chain-termination sequencing the processes of extension, labelling and termination are occurring at the same time. Although the mean size of the oligomers formed can be varied by altering the ddNTP/dNTP ratio, the range spanned is around 450 – 500 bases. More recent sequencing techniques employ a two step-process involving separate labelling and extension stages. In the first labelling step a broad but adjustable range of strand sizes is created by varying the substrate concentration, then the nascent labelled chains are elongated and terminated during the second stage. This can be performed either with the Klenow fragment of DNA polymerase I (13) or by modified T7 DNA polymerase (Sequenase; 14). The use of these methods using either enzyme is outlined in *Table 10* and *11*.

2.2.8 *Troubleshooting poor sequencing reactions*

More artefacts appear to be observed on an autoradiograph after double-stranded DNA sequencing than would be seen if the same insert were sequenced using ssDNA derived from an M13 vector (see Chapter 3). The majority of these artefacts are probably caused by the greater propensity of the denatured DNA to form interchain cross-links through reannealing. Many of the problems are caused through the use of poor quality templates, due to degradation of the sample during preparation or contamination with RNA or other materials, or by storing unprimed denatured DNA for long periods at temperatures where reannealing is likely to occur.

Where compressions are observed in the resulting autoradiograph these can often be eliminated by reducing the formation of hairpin loops in regions of dyad symmetry by using dITP or 7-deaza-dGTP (15) in the sequencing reactions (Chapter 2, Section 3.2). If 7-deaza-dGTP is used it can be substituted directly in the reaction mixture for the dGTP.

Strong bands occurring in all four lanes, often at several points in the sequence ladder, appear to be produced more frequently during double-stranded sequencing. These can often be eliminated by performing the chain extension and termination reactions at higher temperatures to overcome secondary structures. Perhaps the most effective way, however, is to use dITP rather than dGTP and to use Sequenase rather than the Klenow enzyme. If Sequenase is used with dITP, the initial labelling reaction should be performed briefly at room temperature, followed by a rapid extension and termination reaction. A chain-termination mixture for reactions employing Sequenase and dITP is given in *Table 12*. When 7-deaza-dGTP or dITP are used it is important to include a reaction employing the normal mixes on the same gel, as the substitutes may generate artefacts in other regions of the sequence.

If secondary structures are thought to be occurring and the methods described above do not overcome the problem an alternative is to use single-stranded DNA-binding protein in the sequencing reactions. This will prevent secondary structures being formed, but it must be removed by digestion with proteinase K before acrylamide gel separation, as it will otherwise retard the migration of the oligonucleotides. The procedure is also outlined in *Table 12*.

Some of the artefacts that are observed after autoradiography are described in *Table 13*, along with possible causes and remedies. It should be clear from this that the majority of these problems can be avoided by precise adherence to the protocols described above, together with the use of freshly prepared reagents and newly purchased enzymes.

3. GENOMIC SEQUENCING

There are numerous instances where it is desirable to have methods of rapidly determining sequences directly from genomic DNA, circumventing the time-consuming steps involved in cloning the DNA which is being analysed. There is therefore a need for

Table 13. Troubleshooting.

Symptom	Possible Causes	Remedies
Faint bands on developed film.	Loss of DNA on spin-dialysis.	Check recoveries as described in *Table 7*.
	Insufficient primed template.	Increase amounts of DNA or primer; prime for 15 min.
High background on film, with fuzzy bands.	Templates contaminated with RNA.	Check quality of RNase; incubate template in 0.2 M NaOH at 37°C for 15 min before spin-dialysis.
	NaOH passing through spin-dialysis column.	Test efficiency of spin-dialysis as in *Table 7*.
	Template degraded by nucleases.	Process minipreps as rapidly as possible. Phenol extract at the earliest opportunity.
	Template reannealing.	Add primer to samples as quickly as possible; do not prime for longer than 15 min; do not store unprimed templates.
Occasional bands in all four lanes.	Secondary structure in template or synthesized strand.	Perform reactions at 50°C; use 7-deaza-dGTP with the Klenow enzyme, or dITP with Sequenase; use single-stranded DNA binding protein in reactions (see *Table 12* and Section 2.2.8).

techniques to facilitate these analyses. The following sections are concerned with a brief description of the method of Church and Gilbert (16) for direct sequencing of genomic DNA, and the use of the polymerase chain reaction (PCR) to amplify enzymatically a selected DNA fragment (17,18). Current techniques for sequencing the amplified DNA are also discussed.

3.1 Direct sequencing of genomic DNA

The direct genomic sequencing method (16) is limited by the requirement for sequence information to identify specific primer and probe segments in the vicinity of the region to be sequenced, in addition to knowledge of adjacent restriction sites. The method involves the following procedures.

(i)   Restriction of isolated genomic DNA to completion.

(ii)   Random chemical cleavage of the restricted DNA using Maxam−Gilbert methodology (19).

(iii)   Electrophoretic separation of the DNA fragments on a denaturing acrylamide gel.

(iv)   Transfer of the DNA fragments onto a Nylon membrane followed by probing with relatively short single-stranded ^{32}P-labelled probe which is specific for one end of the genomic fragment.

In addition to providing sequence information, this method enables the methylation levels of cytosines in vertebrates and plants to be quantitated, as hydrazine (the reagent

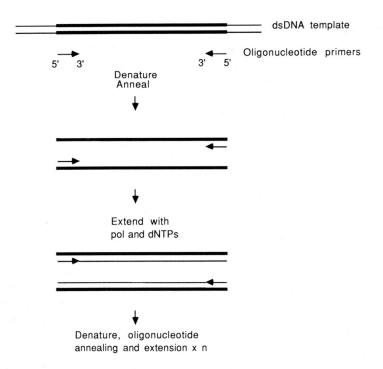

Figure 2. Schematic representation of the PCR. Thickened lines represent the region of DNA to be amplified.

Table 14. Polymerase chain reaction.

1. Mix in a 0.5-ml Eppendorf tube:
    5 μl of 10 $\times$ amplification buffer (500 mM KCl, 100 mM Tris$-$HCl pH 8.4, 25 mM MgCl$_2$, 2 mg ml^{-1} gelatine)
    5 μl of a mix of 5 mM each of dATP, dCTP, dGTP and TTP
    1 μg of genomic DNA
    1$-$2 units Taq pol

    Sterile distilled H$_2$O to 100 μl.

2. Spin briefly in a microcentrifuge and overlay with approximately 100 μl of liquid parafin. Subject the reaction to 25$-$30 cycles of amplification in a programmable heating block,
    1 min at 95°C (denaturation of DNA)
    1 min at 50°C (annealing of primers)
    2 min at 72°C (polymerase extension of primers)

3. Analyse PCR products by agarose or acrylamide gel electrophoresis.

used in Maxam$-$Gilbert sequencing to modify cytosines and thymines prior to cleavage) reacts poorly with 5-Me cytosine.

The limitation in direct genomic sequencing is lack of resolution, however, due to the fact that the DNA of interest is present in extremely low concentrations compared to the level of other DNA.

3.2 **Polymerase chain reaction**

The virtue of the recently developed PCR (17,18) is that it allows the amplification of specific regions of (genomic) DNA by a factor of approximately 10^6. For PCR, suitable primer design relies on knowledge of the sequences associated with the region of DNA to be amplified. Thus the method is of particular use in the diagnosis of genetic disorders (18,20−22), the analysis of allelic sequence variations (23), or any project where the rapid cloning and sequence determination of homologous DNA fragments is required. The enzymatic amplification is directed by sequence-specific primers, and involves repeated cycles of heat denaturation of the DNA, annealing of complementary primers and extension of the annealed primer with a DNA polymerase (*Figure 2*). This results in the exponential increase (2^n) of the target DNA, and over a million copies of the desired DNA sequence may be generated in several hours. Moreover, the error rate over many cycles of amplification is sufficiently low to enable reliable genomic sequence information to be obtained from the amplified product (24).

The earlier reports of the PCR (17,18) involved the use of Klenow fragment of *Escherichia coli* polymerase I. As this enzyme is heat-labile, it was necessary to add fresh polymerase at each cycle. The Klenow polymerase amplification also resulted in the generation of a heterogeneous set of products in addition to the target fragment, probably due to non-specific priming and non-processivity of the enzyme under the conditions of the PCR (23). The recent development of the use of a thermostable polymerase isolated from *Thermus aquaticus* has overcome these problems (25), as synthesis of DNA at a higher temperature allows less non-specific priming. Using this enzyme ('Taq polymerase'), detection of, for example, a single copy of target DNA following amplification from 10 μg of genomic DNA is possible (26). The amplification of RNA transcripts using Taq polymerase can also be carried out using PCR (27); the method involves conversion of the mRNA to a cDNA copy using reverse transcriptase, and then amplification of the cDNA using PCR. Further improvements in the PCR reaction have made the direct amplification of target fragments from genomic DNA routinely possible (26), thus superseding the more time-consuming method involving cDNA synthesis in many instances.

3.2.1 *Equipment and protocol*

Buffers, primers and enzymes for the PCR should be stored at −20°C. It is advisable to store buffers and primers in aliquots to avoid excessive freeze-thawing. Gel purification (28) of oligonucleotide primers is recommended. Programmable Dri-Block PHC-1 may be obtained from Techne (Cambridge) Ltd, Duxford, Cambridge CB2 4PZ, UK. Taq polymerase may be obtained from a variety of commercial suppliers. We have found the Cetus, New England Biolabs and Koch-Light products give satisfactory results.

Reaction. The optimal PCR conditions will depend on the source of polymerase, the DNA to be amplified, and the length of the target sequence and primers. Thus it is not possible to give conditions which are universally appropriate for all PCR reactions, but typically a PCR would be set up as in *Table 14*. Optimization of the reaction conditions for a given amplification is recommended, for example by varying the cycle times and temperatures. The reader is referred to the references cited, and product profiles, for details of suitable conditions.

3.2.2 *Sequencing the amplified products*

The nucleotide sequence of the fragments generated by PCR can be determined in one of several ways. The amplified product can either be cloned into the M13mp sequencing vectors (23,24), or directly sequenced (29,30).

(i) *Cloning into M13 phage.* To facilitate cloning the amplified DNA into M13 (see Chapter 1 for details concerning the use of these vectors), selected restriction sites can be incorporated at the 5' ends of the amplification primers (23,24). Thus amplification followed by restriction and purification of the target DNA generates a fragment which can be readily cloned into M13 RF DNA. The presence of unpaired bases near the 5' ends of the primers does not appear to affect the efficiency of the amplification (23,24), and during later cycles these oligonucleotides anneal to the amplified products with 100% complementarity, rather than the original genomic sequences.

(ii) *Direct sequencing.* Direct sequencing of the amplified DNA is possible using the methodology for sequencing dsDNA in Section 2 of this chapter.

3.2.3 *Generation of ssDNA for sequencing*

A recent development in the use of PCR for genomic sequencing is the generation of ssDNA copies of the target DNA in a single (31−33), or at most, two enzymatic steps (34), thus circumventing the need for M13 cloning. Two methods can currently be used to do this. The more straightforward involves the use of unequal molar amounts of the two amplification primers so that an excess of ssDNA of the selected strand is produced. The ssDNA can be used directly in chain-termination sequencing reactions. In addition, Taq polymerase has been used in chain-termination sequencing reactions containing the ssDNA generated by 'asymmetric PCR' as template (33). The processivity and heat stability of this enzyme make it particularly suitable for sequencing GC-rich DNA, where the formation of secondary structures in the template may cause problems with heat-labile polymerases such as Klenow fragment or Sequenase.

An alternative PCR method involves two steps to generate an RNA copy of the amplified material. This method uses a PCR primer with a phage promoter attached at the 5' end (34). Following amplification, the PCR products are transcribed into RNA and sequenced using reverse transcriptase and chain-termination sequencing (see Chapter 6).

3.3. **Prospects**

Methodology for the sequencing of genomic DNA is developing rapidly. The PCR provides a convenient and rapid way of producing relatively large quantities of sequenceable material from a selected locus in genomic DNA, and enables work which would formerly take weeks to be performed in a several hour automated reaction. The methods for generating ssDNA using PCR will also facilitate the development of automated sequencing systems.

4. REFERENCES

1. Sanger,F., Nicklen,S. and Coulson,A.R. (1977) *Proc. Natl. Acad. Sci. USA,* **74**, 5463.
2. Gronenborn,B. and Messing,J. (1978) *Nature,* **272**, 375.
3. Sanger,F., Coulson,A.R., Barrell,B.G., Smith,A.J.H. and Roe,B.A. (1980) *J. Mol. Biol.,* **143**, 161.

4. Messing,J., Crea,R. and Seeburg,P.H. (1981) *Nucleic Acids Res.*, **9**, 309.
5. Hong,G.F. (1981) *Bioscience Rep.*, **1**, 243.
6. Wallace,R.B., Johnson,M.J., Suggs,S.V., Miyoshi,K., Bhatt,R. and Itakura,K. (1981) *Gene*, **16**, 21.
7. Guo,L., Yang,R.C.A. and Wu,R. (1983) *Nucleic Acids Res.*, **11**, 5521.
8. Chen,E.Y. and Seeburg,P.H. (1985) *DNA*, **4**, 165.
9. Holmes,D.S. and Quigley,M. (1981) *Anal. Biochem.*, **114**, 193.
10. Birnboim,H.C. and Doly,J. (1979) *Nucleic Acids Res.*, **7**, 1513.
11. Keiser,T. (1984) *Plasmid*, **12**, 19.
12. Biggin,M.D., Gibson,T.J. and Hong,G.F. (1983) *Proc. Natl. Acad. Sci. USA*, **80**, 3963.
13. Johnston-Dow,L., Mardis,E., Heiner,C. and Roe,B.A. (1987) *BioTechniques*, **5**, 754.
14. Tabor,S. and Richardson,C.C. (1987) *Proc. Natl. Acad. Sci. USA*, **84**, 4767.
15. Mizusawa,S., Nishimura,S. and Seela,F. (1986) *Nucleic Acids Res.*, **14**, 1319.
16. Church,G.M. and Gilbert,W. (1984) *Proc. Natl. Acad. Sci. USA*, **81**, 1991.
17 Mullis,K. and Faloona,F. (1987) *In Methods in Enzymology*. Wu,R. (ed.), Academic Press, New York, Vol. **155**, p.335.
18. Saiki,R.K., Scharf,S., Faloona,F., Mullis,K., Horn,G.T., Erlich,H.A. and Arnheim,N. (1985) *Science*, **230**, 1350.
19. Maxam,A.M. and Gilbert,W. (1977) *Proc. Natl. Acad. Sci. USA*, **74**, 560.
20. Saiki,R.K., Bugawan,T.L., Horn,G.T., Mullis,K.B. and Erlich,H.A. (1986) *Nature*, **324**, 163.
21. Embury,S.H., Scharf,S.J., Saiki,R.K., Gholson,M.A., Golbus,M. Arnheim,N. and Erlich,H.A. (1987) *New Eng. J. Med.*, **316**, 656.
22. Kogan,S.C., Doherty,M. and Gitschier,J. (1987) *New Engl. J. Med.*, **317**, 985.
23. Horn,G.T., Bugawan,T.L., Long,C.M. and Erlich,H. (1988) *Proc. Natl. Acad. Sci. USA*, **85**, 6012.
24. Scharf,S.J., Horn,G.T. and Erlich,H. (1986) *Science*, **233**, 1076.
25. Chehab,F.F., Doherty,M., Cai,S., Kan,Y.W., Cooper,S. and Rubin,E.M. (1987) *Nature*, **329**, 293.
26. Saiki,R.K., Gelfand,D.H., Stoffel,S., Scharf,S.J., Higuchi,R., Horn,G.T., Mullis,K.B. and Erlich,H.A. (1988) *Science*, **239**, 487.
27. Powell,L.M., Wallis,S.C., Pease,R.J., Edwards,Y.H., Knott,T.J. and Scott,J. (1987) *Cell*, **50**, 831.
28. Carter,P., Bedouille,H., Waye,M.M.Y. and Winter,G. (1985) In *Oligonucleotide Site-directed Mutagenesis in M13*. Anglian Biotechnology Ltd, Hawkins Road, Colchester, Essex CO2 8JX, UK.
29. Wong,C., Dowling,C.E., Saiki,R.K., Higuchi,R.G., Erlich,H.A. and Kazazian,H.H. (1987) *Nature*, **330**, 384.
30. Wrishnik,J.A., Higuchi,R.G., Stoneking,M., Erlich,H.A., Arnheim,N. and Wilson,A.C. (1987) *Nucleic Acids Res.*, **15**, 529.
31. Gyllenstein,U.B. and Erlich,H.A. (1988) *Proc. Natl. Acad. Sci. USA*, **85**, 7652.
32. Paabo,S., Gifford,J.A. and Wilson,A.C. (1988) *Nucleic Acids Res.*, **16**, 9775.
33. Innis,M.A., Myambo,K.B., Gelfand,D.H. and Brow,M.A.D. (1988) *Proc. Natl. Acad. Sci. USA*, **85**, 9436.
34. Stoflet,E.S., Koeberl,D.D., Sarkar,G. and Sommer,S.S. (1988) *Science*, **239**, 491.

CHAPTER 5

Maxam and Gilbert sequencing using one metre gel systems

R.F.BARKER

1. INTRODUCTION

The original procedure for chemical sequencing of deoxyribonucleic acid was first published by Allan M.Maxam and Walter Gilbert in 1977 (1). They developed a new technique for sequencing DNA by a chemical procedure that breaks a terminally-labelled DNA molecule partially at each repetition of a base. The lengths of the labelled fragments then identify the positions of that base. They described reactions that cleave DNA preferentially at guanines, at adenines and guanines equally, at cytosines and thymines equally, and at cytosines alone. When the products of these four reactions are resolved by size, by electrophoresis on a polyacrylamide gel, the DNA sequence can be read from the pattern of radioactive bands. The technique permitted the sequencing of at least 100 bases from the point of labelling.

This procedure was described in more detail in 1980 (2), and this publication has become the basic protocol followed by research workers employing this technique to sequence DNA. However, many modifications to the original protocol have been made over the years to optimize the sequencing reactions and improve the resolution of the end-labelled DNA fragments. Together with the development of one metre long sequencing gel systems (3) the number of nucleotides that can be read at any one time can be increased to over 600.

Further modifications to the protocol have been made to reduce the time taken in end-labelling and sequencing the DNA and to substitute the hazardous chemicals employed with potentially less harmful ones. The Maxam and Gilbert method of DNA sequencing is still preferred by many researchers throughout the world to the alternative dideoxy chain-termination method of Sanger (4,5). It has some distinct advantages over the Sanger method and these will be discussed later. In this chapter I will attempt to outline a method of sequencing DNA using the Maxam and Gilbert methodology adapted to one metre long gel systems.

The advantage of running one metre gel systems is in the increased resolution produced when sequencing large DNA fragments. The initial 300 bp from the end-labelled end of a DNA fragment are relatively easy to resolve; however, large oligonucleotides are much more difficult to separate. A one metre 4% polyacrylamide gel routinely produces sequences starting 300 bp from the end-labelled end and continuing out to 600 bp. Longer sequences of up to 750 bp have been recorded.

2. SELECTION OF A SUITABLE SEQUENCING VECTOR

Prior to sequencing any DNA fragment it is important to select a suitable vector into which to clone the target DNA. The most commonly used are the pUC series of cloning vectors derived by Messing and co-workers (6,7, and see Chapter 1). These are ideal for Maxam and Gilbert sequencing as they are small in size (~ 2700 bp), and contain multiple restriction endonuclease cleavage sites, in either orientation relative to the *lacZ* gene, which can be utilized in the sequencing strategy.

There are also vectors available which are particularly suitable for Maxam and Gilbert sequencing such as pUR222. DNA fragments cloned into this plasmid can be sequenced without isolation of labelled fragments (8).

The most important fact to consider when selecting a vector is its size. It is possible to sequence directly from lambda clones of over 50 kb in length but it becomes difficult to select the correct restriction fragments for sequencing from the many produced.

3. PREPARATION OF DNA SAMPLES PRIOR TO MAXAM AND GILBERT SEQUENCING

In order to obtain unambiguous results great care must be taken when preparing any DNA sample for sequencing. The DNA must be isolated from a culture grown from a single colony of transformed cells and be completely free of contaminating host cell chromosomal DNA and RNA. Any contaminating nucleic acids will decrease the efficiency of end-labelling of the required DNA and cause confusion when isolating the correct DNA fragments for the subsequent sequencing reactions. The following procedure provides $500-1000$ μg of purified plasmid DNA suitable for sequencing.

3.1 Large scale plasmid preparation

(i)   Inoculate 50 ml of L broth (10 g of Bactotryptone, 5 g of yeast extract, 10 g of NaCl; dissolve, adjust pH to 7.5 with NaOH, make up to 1 litre and autoclave) with a single colony of transformed bacteria. Incubate the culture overnight at 37°C with shaking.

(ii)   Dilute 2 ml of the overnight culture into one litre of L broth and grow the bacteria at 37°C to an OD_{600} of approximately 0.5.

(iii)   Add 175 mg of chloramphenicol (final concentration of 175 μg ml^{-1}).

(iv)   Incubate the cultures at 37°C for 16 h with shaking.

(v)   Harvest the bacteria by centrifugation at 4000 g for 5 min and drain the resultant pellets.

(vi)   Resuspend the pellets in a total volume of 36 ml of 50 mM glucose, 25 mM Tris$-$HCl pH 8, 10 mM Na_2EDTA.

(vii)   Add 4 ml of a 40 mg ml^{-1} solution of lysozyme in the buffer indicated in (vi) to the suspension. Incubate the mixture at room temperature for 10 min. Cool on ice.

(viii) Add 80 ml (2 vol) of freshly made 0.2 M sodium hydroxide, 1% (w/v) SDS and swirl this mixture gently. (Do not shake too vigorously as this will release chromosomal DNA.) Let lysate stand on ice for 10 min.

(ix)   Add 40 ml (1 vol) of ice-cold 5 M acetate solution (prepared by making a solution of 3 M with respect to potassium acetate and 2 M with respect to acetic acid) to the suspension. Mix well, and a white precipitate should form. Incubate the mixture on ice for 15 min.

(x) Collect the precipitate by centrifugation at 24 000 *g* for 15 min and filter the supernatant through 4−5 layers of cheesecloth.

(xi) Add 0.6 vol of isopropanol to the filtrate, mix thoroughly and incubate for 10 min at room temperature.

(xii) Collect the resultant precipitate by centrifugation at 12 000 *g* for 10 min.

(xiii) Wash the pellets with 70% ethanol containing 200 mM Tris−HCl pH 8.0, recentrifuge as above and drain.

(xiv) Resuspend the pellets in a total volume of 3.85 ml of 10 mM Tris−HCl, 1 mM Na_2EDTA, pH 7.5 (TE).

(xv) For every 3.85 ml of TE solution add 4.4 g of caesium chloride.

(xvi) Add 320 μl of a 10 mg ml^{-1} solution of ethidium bromide in TE. Adjust the refractive index to 1.391 at 25°C by the addition of either TE or caesium chloride (if necessary).

(xvii) Spin at 45 000 r.p.m. for 16 h in a high speed centrifuge at 20°C.

(xviii) The gradients generally contain three bands; the lowest band does not bind ethidium bromide; both of the upper two bands bind ethidium bromide but on many gradients the less dense band of the two, which corresponds to chromosomal DNA, can barely be seen.

(xix) Collect the plasmid band, which corresponds to the lower of the two bands which bind ethidium bromide, by puncturing the side of the tube below the band with a 20 gauge needle and allowing the plasmid band to flow through the needle into a collecting tube.

(xx) Transfer the recovered plasmid into a quick seal tube that is then filled with a solution made by dissolving 98.2 g of caesium chloride in 100 ml of TE and 2.73 ml of 10 mg ml^{-1} ethidium bromide.

(xxi) Spin for 36 h at 47 000 *g* in a high speed centrifuge at 20°C.

(xxii) Collect the plasmid band as before, and remove the ethidium bromide by extraction with isopropanol which has been saturated by adding 10 ml of 5 M sodium chloride and 10 ml of 50 mM Tris−HCl, 1 mM Na_2EDTA pH 7.5 to 80 ml of isopropanol. The extracted plasmid is diluted twice with TE pH 7.5 and precipitated with 2 vol of ethanol at −20°C.

3.2 Oligonucleotides

Synthetic oligonucleotides are now readily synthesized on commercially available equipment. The oligonucleotides when cleaved from the solid supports are lacking the naturally occurring 5'-phosphate group so that they are immediately ready for 5' end-labelling prior to the sequencing reactions.

It is necessary, however to purify the final product either by HPLC or gel electrophoresis techniques before the oligonucleotides can be successfully sequenced.

4. END-LABELLING DNA FRAGMENTS

4.1 5' end-labelling

The procedure for 5' end-labelling of native DNA molecules and synthetic

Table 1. 10 × Universal restriction enzyme buffer.

Stock solutions	10 × buffer solution (ml)	Final concentration (mM)
1 M Tris−acetic acid pH 7.8	3.3	330
5 M potassium acetate	1.3	650
1 M magnesium acetate	1.0	100
0.1 M spermidine	4.0	40
0.1 M dithiothreitol	0.5	5
	10.1	

oligonucleotides is outlined below. The method has proved to be very efficient for labelling synthetic oligonucleotides and also for native DNA which has been cut with restriction enzymes leaving a 5′ overhang. Restriction enzymes leaving blunt ends or 5′ recessed ends on the DNA can also be end-labelled using this procedure if a denaturing step is employed prior to the kinasing reaction.

4.1.1 Initial cutting of plasmid DNA

(i)   Add 10 μg of purified plasmid DNA to 5 μl of 10 × restriction enzyme buffer (see *Table 1*).

(ii) Add 20 units of restriction enzyme and make up to 50 μl final volume with double distilled sterile water.

(iii) Incubate the reaction mixture for 1 h at 37°C.

It is important to check this initial digestion for completion by running a sample on an agarose gel. If the digestion is complete proceed to the next step. If the digestion is incomplete the end-labelling efficiency will be reduced and partial bands will give rise to unexpected end-labelled fragments. If incomplete, add more enzyme to the sample tube and incubate for a further hour. Recheck the digest again after this time.

4.1.2 Phosphatase reaction

The naturally occurring 5′ phosphate group is removed by the action of the enzyme alkaline phosphatase in the following reaction.

(i)   Add 5 μl or 1/10 vol of 1 M Tris−HCl pH 8.4 to the sample tube.

(ii) Add 5 μl (5 units) of calf intestinal alkaline phosphatase (Boehringer, Mannheim), which is prepared as follows.

(a) Dissolve the contents of 1 vial of lyophilized alkaline phosphatase (4000 units) in 2 ml of water containing 80 μl of 1 M Tris−HCl pH 8.0, and 4 μl of 1 M MgCl$_2$.

(b) Add 2.0 ml of sterile 80% glycerol.

(c) Mix well and store at −20°C.

(iii) Incubate the sample for 30 min at 55°C.

4.1.3 Phenol extraction

Phenol extraction at this stage is necessary to remove all traces of the alkaline phosphatase prior to the kinasing reaction. Equilibrated phenol is prepared in the following way.

(i)   Melt redistilled phenol in a 65°C water bath.

(ii)   Add an equal volume of 1 × TE buffer (100 mM Tris−HCl pH 8.0, 10 mM Na$_2$EDTA).

(iii) Add 2 M Tris−HCl pH 9.5 (~65 ml) until the phenol has a pH of between 7 and 8.

(iv)   Add 0.1% (w/v) 8-hydroxyquinoline (optional).

(v)   Store at 4°C.

The procedure for phenol extraction is as follows.

(i)   Add 200 μl of equilibrated phenol to the sample and mix for 1 min on a vortex mixer. Spin for 2 min at 12 000 g at room temperature.

(ii)   Remove the top aqueous layer into a clean 1.5-ml Eppendorf tube.

(iii) Add to the aqueous layer 100 μl of equilibrated phenol and 100 μl of chloroform: isoamyl alcohol, 25:1 (v/v). Mix for 1 min and spin for 1 min at 12 000 g at room temperature.

(iv)   Remove the top aqueous layer into a clean 1.5-ml Eppendorf tube.

(v)   Add to the aqueous layer 200 μl of chloroform:isoamyl alcohol, 25:1 (v/v). Mix for 30 sec and spin for 30 sec at 12 000 g at room temperature.

(vi)   Remove the top aqueous layer into a clean 1.5-ml Eppendorf tube.

4.1.4 *Ethanol precipitation*

(i)   Add to the aqueous layer 5 μl of 3.0 M sodium acetate pH 6.0 and mix.

(ii)   Add 1.0 ml of −20°C 95% ethanol and mix.

(iii) Spin for 15 min at 12 000 g at 4°C.

(iv)   Remove the supernatant and wash the DNA pellet with 800 μl of −20°C 95% ethanol.

(v)   Spin for 10 min at 12 000 g at 4°C.

(vi)   Remove the supernatant and dry the DNA pellet under vacuum.

If the sample DNA was initially cut with a restriction enzyme producing blunt ends or a recessed 5′ end it will now be necessary to denature the DNA to enable a radioactive phosphate group to be incorporated by polynucleotide kinase at the 5′ ends of the DNA molecules.

4.1.5 *Denaturation*

(i)   Add to the dried DNA 15 μl of water, mix to dissolve.

(ii)   Add 15 μl of denaturation buffer (20 mM Tris−HCl pH 9.5, 0.1 mM Na$_2$EDTA, 1.0 mM spermidine).

(iii) Incubate the sample for 5 min at 70°C and then place on ice.

4.1.6 *Kinasing reaction*

(i)   Add to the denatured sample 4.0 μl of kinase buffer (500 mM Tris−HCl pH 9.5, 100 mM MgCl$_2$, 50 mM dithiothreitol in 50% glycerol).

(ii)   Add 2 μl of T4 polynucleotide kinase (5 units per μl, Amersham).

(iii) Add 10 μl of aqueous [γ-^{32}P]adenosine 5′ triphosphate solution [2 mCi ml^{-1}, 3000 Ci mmol^{-1}, Dupont (UK) Ltd].

(iv)   Mix and incubate for 30 min at 37°C.

(v)   Ethanol-precipitate, carefully removing the supernatant whilst monitoring with a hand monitor. (Most of the label should remain with the precipitate.) Dry the DNA pellet under vacuum.

The DNA fragments should now be 5′ end-labelled at both ends of the double-stranded DNA. It is now necessary to produce DNA fragments labelled at only one end. This can be achieved either by strand separation or by recutting the fragments with a second restriction enzyme.

4.1.7 *Strand separation*

(i)   Mix 1 μg of the ^{32}P-labelled DNA with 40 μl of 30% (v/v) dimethyl sulphoxide, 1 mM Na$_2$EDTA, 0.05% (w/v) bromophenol blue and 0.05% (w/v) xylene cyanol.

(ii)   Incubate for 2 min at 90°C and quick-chill in iced water.

(iii)   Immediately load onto a pre-electrophoresed (300 V for 2 h) denaturing polyacrylamide gel containing 5.0% (w/v) acrylamide, 0.1% (w/v) *N,N′*-methylene-bis-acrylamide, 5% (w/v) urea, 50 mM Tris−borate, pH 8.3, 1 mM Na$_2$EDTA. For 100 ml of gel mix, add 1 ml of 10% (w/v) ammonium persulphate and 200 μl of TEMED to polymerize.

Use a vertical slab gel 1.5 × 200 × 400 mm and electrophorese at 300 V (8 V cm^{-1}). It is important that the gel does not heat up. Stop the gel when the marker dyes have migrated to the desired positions according to the size of the fragments being analysed (the bromophenol blue and xylene cyanol co-migrate approximately with DNA fragments of 100 bases and 200 bases respectively. The bromophenol blue is routinely run to the bottom of the gel).

4.1.8 *Secondary restriction enzyme digestion*

(i)   Add to the dried end-labelled DNA 40 μl of double distilled sterile water, mix to dissolve.

(ii)   Add 5 μl of 10 × restriction enzyme buffer (see *Table 1*).

(iii)   Add 20 units of restriction enzyme, mix and incubate for 2 h at 37°C.

This second digest can be checked on a 0.8% agarose gel for complete digestion. However, the longer incubation time should provide complete digestion of the DNA. Omitting the second gel also reduces the possible contamination when handling radioactively-labelled samples.

(iv)   Add 10 μl of 50% glycerol, 25 mM Na$_2$EDTA, 0.25% (w/v) bromophenol blue, 0.25% (w/v) xylene cyanol.

(v)   Load onto a non-denaturing polyacrylamide eluting gel.

4.1.9 *Polyacrylamide eluting gel (non-denaturing)*

(i)   Stock acrylamide solution: 48 g of acrylamide and 2 g of *N,N′*-methylene-bis-acrylamide. Make up to 100 ml with double distilled sterile water.

(ii)   10 × TBE buffer: 107.8 g of Tris base, 55 g of boric acid, 9.3 g of Na$_2$EDTA, dissolved in 1 litre of water.

(iii)   Gel solution: 15 ml of stock acrylamide solution, 10 ml of 10 × TBE buffer,

25 ml of glycerol and 50 ml of water. Polymerize by addition of 1 ml of 10% (w/v) ammonium persulphate and 200 μl of TEMED.

Use a vertical slab gel 1.5 $\times$ 200 $\times$ 400 mm and electrophorese at 400 V (10 V cm^{-1}). Run the gel until the marker dyes have migrated to the desired positions according to the size of the fragments being separated (the bromophenol blue and xylene cyanol co-migrate approximately with DNA fragments of 100 base pairs and 200 base pairs respectively. The bromophenol blue is routinely run to the bottom of the gel). Fragments shorter than 2000 bp can be separated using this method. For longer fragments agarose gels have to be employed.

4.1.10 *Autoradiography*

After electrophoresis of the radioactively-labelled fragments the gels are dismantled and one of the glass plates removed. The gel is then covered with 'Saran Wrap'. Radioactive ink is then spotted onto thin strips of 3MM filter paper and placed asymmetrically at the four corners of the gel and fixed with clear tape. A piece of X-ray film is laid on the gel, weighted down by another glass plate. Autoradiography for 2 min at room temperature is usually sufficient.

The X-ray film is then developed and the fragments to be sequenced are identified by their size and mobility. *Figure 1* shows a typical autoradiograph of 5' end-labelled DNA samples. When using restriction enzymes with a six base recognition sequence for the initial cut there are relatively few bands generated as shown in lanes 1, 2, 3, 7 and 8. It is convenient to run adjacent samples utilizing the same restriction enzymes in reverse roles for the initial and secondary cuts (Sections 4.1.1 and 4.1.8) as shown in lane 1 with an initial *Eco*RI cut, followed by end-labelling and then recutting with *Nco*I, and lane 2 with an initial *Nco*I cut, recut with *Eco*RI. Bands migrating with the same mobility then correspond to the same fragment end-labelled only at one end either from the *Eco*RI site in lane 1 band A, or the *Nco*I site in lane 2 band B. Single bands appearing in only one lane are therefore double end-labelled fragments, e.g. lane 2 band C.

When using restriction enzymes with a four or five base recognition site for the initial cut many bands are produced, as shown in lanes 4, 5 and 6. Using two enzymes in reverse roles in adjacent lanes, bands of corresponding mobility can be isolated; lane 4 *Dde*I−*Hin*fI band D and lane 5 *Hin*fI−*Dde*I band E. This method can be used for shotgun sequencing strategies of DNA using several combinations of restriction enzymes with four or five base recognition sites. Specific bands can be isolated from restriction enzymes with four or five base recognition sites when used in conjunction with a six base cutter. Lane 6 shows an initial cut with *Dde*I recut with *Xba*I and lane 7 shows an initial cut with *Xba*I recut with *Dde*I. The bands in lane 6, F and G, are adjacent to the bands in lane 7 H and I respectively and correspond to single end-labelled *Dde*I to *Xba*I fragments. The sample in lane 8 was initially digested with *Sph*I, which produces a 3' overhang. The sample was then denatured before the kinase reaction and recut with *Nco*I. The 5' end-labelling is not as efficient as samples 1−7 but there is sufficient activity to be able to sequence both visible bands.

(i)   Cut windows in the X-ray film at the position of the fragments required, and use the radioactive ink spots to align the X-ray film on the eluting gel.

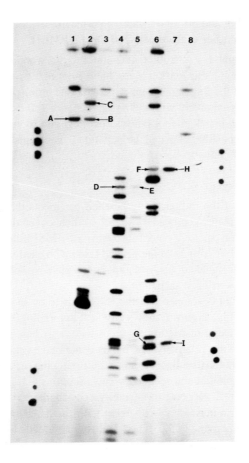

Figure 1. An autoradiograph (exposed for 2 min) of DNA samples 5′ end-labelled with ^{32}P and run on a 7.5% polyacrylamide gel. **Lanes 1−8** show initial and secondary restriction enzyme digests in the following order: **1**, *Eco*RI−*Nco*I; **2**, *Nco*I−*Eco*RI; **3**, *Bam*HI−*Nco*I; **4**, *Dde*I−*Hin*fI; **5**, *Hin*fI−*Dde*I; **6**, *Dde*I−*Xba*I; **7**, *Xba*I−*Dde*I; **8**, *Sph*I−*Nco*I. The bands indicated by the arrows and lettered A to I are referred to in the text.

(ii)   Using a scalpel remove the gel through the window cut in the X-ray film. These gel fragments should correspond to the single end-labelled fragments of DNA. Check the number of counts on a Geiger hand monitor, which should be greater than 2000 c.p.s.

5. ELUTION OF SAMPLES

The two most commonly used methods are mechanical and electroelution. There are several commercial electroelution devices available today which provide high levels of recovery; however, the method in Section 5.2 gives recovery yields of greater than 90%.

5.1 **Mechanical elution**

(i)   Place the gel slice in a 1.5-ml Eppendorf tube.

(ii)   Crush the slice with a blunt-ended glass rod.

(iii)   Add 400 μl of elution solution (500 mM ammonium acetate, 10 mM magnesium acetate, 1 mM Na$_2$EDTA, 0.1% (w/v) SDS).

(iv)   Close the lid tightly and seal with parafilm.

(v)   Shake the sample at 37°C for a minimum of 4 h or overnight.

(vi)   Plug a blue pipette tip with siliconized glass wool.

(vii)   Place all the sample into the pipette tip and allow it to filter into an Eppendorf tube.

(viii) When all the sample has passed through add a further 100 μl of elution solution to wash the retained crushed gel.

(ix)   Check with a hand monitor that the filtrate contains the majority of counts.

(x)   Precipitate the DNA by adding 1.0 ml of 95% ethanol at -20°C. Mix well and spin at 15 000 g for 15 min at 4°C.

(xi)   Redissolve the precipitate in 200 μl of 0.3 M sodium acetate pH 6.0. Precipitate the DNA by adding 800 μl of 95% ethanol at -20°C. Mix well and spin at 15 000 g for 15 min at 4°C.

(xii)   Wash the pellet with 800 μl of 95% ethanol at -20°C and dry under vacuum.

The sample is now ready for the Maxam and Gilbert sequencing reactions (Section 7).

5.2 **Electroelution**

(i)   Place the gel slice into a small length of 1 cm diameter dialysis tubing clamped at one end.

(ii)   Add 1 × TBE buffer to fill the tubing, removing all air bubbles.

(iii)   Extrude excess buffer from the tubing and clamp the other end ensuring that the clamps are close to the gel slice but not crushing it.

(iv)   Place the sample on the flat gel bed of any horizontal electrophoresis apparatus and cover with 1 × TBE buffer.

(v)   Apply a voltage across the apparatus; usually a constant 100 V.

(vi)   Run until the sample has migrated from the gel and bound to the dialysis membrane. This can be visualized if the fragment is stained with ethidium bromide. If not stained, note that the time required depends on the size of the DNA fragment and the composition of the gel. Electrophoresis for 1–2 h at 100 V will elute fragments of up to 50 kb.

(vii)   Reverse the current for 30 sec to remove the sample from the dialysis membrane.

(viii) Using a fine pipette tip, remove the buffer from the dialysis tubing and place in an Eppendorf tube.

(ix)   Wash out the dialysis tube with one half volume of 1 × TBE buffer, and add to the sample.

(x)   Phenol-extract the sample and ethanol-precipitate the DNA.

6. 5′ END-LABELLING SYNTHETIC DNA OLIGONUCLEOTIDES

Synthetic oligonucleotides are 5′ end-labelled prior to Maxam and Gilbert sequencing using the following reaction conditions.

(i)   To 100 ng of oligonucleotide in a volume of 20 μl of double distilled sterile water add: 1 μl of T4 polynucleotide kinase (5 units μl^{-1}), 2 μl of kinase buffer (Section 4.1.6), 1 μl of [γ-^{32}P]adenosine 5' triphosphate (2 mCi ml^{-1}, 3000 Ci mmol^{-1}).

(ii)   Mix and incubate for 30 min at 37°C.

(iii)   Add 10 μl of 50% glycerol, 25 mM Na$_2$EDTA, 0.25% (w/v) xylene cyanol loading dye.

(iv)   Load all of the sample onto a 20% non-denaturing polyacrylamide gel 1.5 mm $\times$ 200 mm $\times$ 400 mm and electrophorese at 300 V (8 V per cm) until the dye has migrated the desired distance (the bromophenol blue and xylene cyanol co-migrate approximately with DNA fragments of 100 bases and 200 bases respectively. The bromophenol blue is routinely run to the bottom of the gel).

(v)   Carry out autoradiography as in Section 4.1.10 and elute the sample as in Section 5.

The samples are now ready for the Maxam and Gilbert sequencing reactions.

7. MAXAM AND GILBERT SEQUENCING REACTIONS FOR LONG DOUBLE- OR SINGLE-STRANDED DNA MOLECULES

(i)   Dissolve the dried single end-labelled DNA fragments in 26 μl of double distilled sterile water.

(ii)   Add 4 μl of carrier DNA at a concentration of 1 μg per μl.

(iii)   Label four Eppendorf tubes G, A, T, C.

7.1 **G reaction**

(i)   Place the G-labelled Eppendorf tubes on ice.

(ii)   Add 200 μl of DMS buffer (50 mM sodium cacodylate, pH 8.0, 1 mM Na$_2$EDTA).

(iii)   Add 5 μl of the end-labelled DNA mix and keep on ice.

(iv)   Add 1.0 μl of dimethyl sulphate.

(v)   Mix and incubate for 20 sec at 20°C.

(vi)   Add 50 μl of DMS stop solution (1.5 M sodium acetate pH 7.0, 1.0 M mercaptoethanol, 100 μg ml^{-1} tRNA) and 800 μl of 95% ethanol at -20°C. Mix well and spin at 12 000 g for 15 min at 4°C.

(vii)   Withdraw the supernatant checking with a hand monitor that no end-labelled DNA is removed.

(viii)   Redissolve the DNA in 400 μl 0.3 M sodium acetate pH 6.0.

(ix)   Precipitate the DNA by adding 800 μl of 95% ethanol at -20°C. Mix and spin at 12 000 g for 15 min at 4°C.

(x)   Discard the supernatant and wash the DNA pellet with 800 μl of 95% ethanol.

(xi)   Spin at 12 000 g for 10 min at 4°C and discard the supernatant.

(xii)   Dry under vacuum for 15 min.

7.2 **G+A reaction**

(i)   Place the A-labelled Eppendorf tubes on ice.

(ii)   Add 15 μl of double distilled sterile water.

(iii)   Add 10 μl of the end-labelled DNA mix and keep on ice.

(iv)   Add 30 μl of formic acid.

(v)   Mix and incubate at 20°C for 3 min.

(vi)   Stop the reaction by adding 400 μl of hydrazine stop (0.3 M sodium acetate, 0.1 mM Na$_2$EDTA, 25 μg ml^{-1} tRNA) and 800 μl of 95% ethanol at −20°C. Mix well and spin at 12 000 g for 15 min at 4°C.

(vii) Follow steps (vii) to (xii) in Section 7.1 above.

7.3 T+C reaction

(i)   Place T-labelled Eppendorf tubes on ice.

(ii)   Add 15 μl of double distilled sterile water.

(iii) Add 10 μl of end-labelled DNA mix and keep on ice.

(iv)   Add 30 μl of hydrazine.

(v)   Mix and incubate for 3 min at 20°C.

(vi)   Stop the reaction by adding 400 μl of hydrazine stop and 800 μl of 95% ethanol at −20°C. Mix well and spin at 12 000 g for 15 min at 4°C.

(vii) Follow steps (vii) to (xii) in Section 7.1 above.

7.4 C reaction

(i)   Place C-labelled Eppendorf tubes on ice.

(ii)   Add 15 μl of 5.0 M sodium chloride.

(iii) Add 5 μl of end-labelled DNA, mix and keep on ice.

(iv)   Add 30 μl of hydrazine.

(v)   Mix and incubate for 3 min at 20°C.

(vi)   Stop the reaction by adding 400 μl of hydrazine stop and 800 μl of 95% ethanol at −20°C. Mix well and spin at 12 000 g for 15 min at 4°C.

(vii) Follow steps (vii) to (xii) in Section 7.1 above.

All the dried samples are now treated with piperidine which will remove the modified bases and cleave the DNA.

7.5 Piperidine reaction

(i)   Prepare a fresh solution of 1.0 M piperidine.

(ii)   Add 100 μl of 1.0 M piperidine to the dried DNA samples and mix well.

(iii) Make sure that the caps of the tubes are securely shut and pierce three holes in the lid with a gauge 20 syringe needle.

(iv)   Incubate the tubes at 90°C for 30 min. Cool on ice.

(v)   Add 10 μl of 3.0 M sodium acetate pH 6.0 and mix.

(vi)   Precipitate the DNA by adding 800 μl of 95% ethanol at −20°C and mixing. Spin at 12 000 g for 15 min at 4°C.

(vii) Discard the supernatant and redissolve the DNA pellet in 200 μl of 0.3 M sodium acetate pH 6.0.

(viii) Precipitate the DNA by adding 800 μl of 95% ethanol at −20°C and mixing. Spin at 12 000 g for 15 min at 4°C.

(ix)   Discard the supernatant and wash the DNA pellet with 800 μl of 95% ethanol at −20°C. Spin at 12 000 g for 15 min at 4°C.

(x)   Discard the supernatant and dry the DNA pellet under vacuum.

Table 2. Determination of autoradiography times.

Cerenkov counts (c.p.m.)	Amount of loading dye (µl)	Autoradiography time (days)
10 000	6	7
10 – 50 000	8	5
50 – 100 000	10	1
100 – 200 000	15	0.5

The amount of end-labelled DNA can now be accurately determined by measuring Cerenkov counts in a scintillation counter. This will determine the amount of loading dye necessary to add to produce an even number of counts in the samples and also the autoradiography exposure time. See *Table 2*.

8. MAXAM AND GILBERT SEQUENCING REACTIONS FOR SYNTHETIC DNA OLIGONUCLEOTIDES

Short synthetic DNA oligonucleotides of less than 100 bases are sequenced using the same methods as described in Section 7. The only alterations are in the times of the base modification reactions. The times of these are increased accordingly.

(i)   G reaction to 20 min.

(ii)   G+A reaction to 1 h.

(iii)   T+C reaction to 1 h.

(iv)   C reaction to 1 h.

(v)   The cleavage products are then separated on a 20% denaturing polyacrylamide gel.

9. ELECTROPHORESIS

To obtain the optimum sequencing data from an end-labelled fragment of DNA a thermostatically controlled ultrathin 1 metre long gel is employed.

9.1 Electrophoresis apparatus

There are several 1 metre gel systems available commercially. One designed by Sequencing Systems Ltd in Cambridge exhibits many beneficial features in safety and operation. The system as shown in *Figure 2* is completely enclosed in a safety cabinet with low voltage interlocks on the lid and door. All the wiring and connectors are rated at 5000 V. The upper and lower buffer chambers are removable for cleaning purposes. The 1 metre long thermostatting plate is heated by an enclosed electrical element controlled by a thermostat. A control box is supplied, where the required temperature of the plate can be set and the actual temperature of the plate is displayed. The complete system is available from Koch-Light Ltd, Haverhill, Suffolk.

9.2 Treatment of the gel plates

When pouring ultrathin polyacrylamide gels it is essential that the glass plates used are of the highest quality plate glass, ensuring that the gel will be of uniform thickness throughout its length. It is also imperative that the plates are adequately cleaned as this will greatly reduce the occurrence of air bubbles when the gel is poured. It is essential

Figure 2. A one metre electrophoresis gel system specifically designed for DNA sequencing. The temperature of the thermostatting plate which has an embedded heating element is controlled exactly by means of a thermostatting unit (not shown). Each thermostatting control box maintains two gel rigs. The system is available from Koch-Light Ltd, Haverhill, Suffolk.

that after electrophoresis the polyacrylamide gel does not bind to the thermostatting plate but binds firmly to the gel plate. A method for chemically binding acrylamide to glass plates was described by Garoff and Ansorge (9), and this method is described below.

9.2.1 *Thermostatting plate*

(i)   Wash the plate with a mild detergent and water.

(ii)   Wash twice with distilled water to remove all traces of the detergent.

(iii)   Wash twice with ethanol.

(iv)   Apply a liberal amount of dimethyldichlorosilane solution ('Repelcote', BDH) making sure that it is distributed evenly over the plate using a lint-free tissue. It is advisable to carry out this procedure in a fume hood.

(v)   Allow the plate to dry for 2 min.

(vi)   Polish the plate with lint-free tissues.

(vii)   Repeat the treatment with the 'Repelcote' to ensure that the whole surface area of the plate is covered.

(viii)   Wash once with ethanol with a lint-free tissue.

9.2.2 *Gel plate*

(i)   Wash the plate with a mild detergent and distilled water.

(ii)   Wash twice with distilled water to remove all traces of detergent.

(iii)   Wash twice with ethanol.

(iv)   Prepare a fresh solution consisting of 25 ml of ethanol, 0.75 ml of 10% acetic acid and 75 μl of 3-methacryloxypropyltrimethoxy silane.

(v)   Pour this solution onto the gel plate and distribute evenly over the surface with a lint-free tissue.

(vi)   Allow to dry for 2 min.

(vii)   Wash liberally with ethanol to remove all excess silane. The number of ethanol washes will depend on the concentration of polyacrylamide gel being used. For a 4% gel wash the plate four times, for a 6% gel wash the plate twice and for a 16% gel simply polish the plate with a lint-free tissue.

It is important to remove the excess silane, as it will diffuse through the gel causing the gel to bind to both the gel and thermostatting plates.

9.3 **Pouring an ultrathin gel**

(i)   Place the thermostatting plate on a flat surface.

(ii)   Lay 110 cm long, 1 cm wide, 0.2 mm thick Teflon spacers along the edges of the plate.

(iii)   Place the gel plate, with the treated surface facing the thermostatting plate, on top of the Teflon spacers.

(iv)   Clamp the edges of the plates together using clips, approximately eight per side.

(v)   Use 100 ml of a denaturing polyacrylamide gel solution. The concentration of polyacrylamide will depend on the application required (see *Table 3*, Section 9.5). To obtain the maximum number of bases per fragment a low percentage gel is required, usually 4%. For shorter readings higher percentage gels should be prepared. For sequencing oligonucleotides a 20% gel may be necessary.

	Acrylamide stock solutions			
	4%	6%	16%	20%
Urea	500 g	500 g	500 g	500 g
N,N'-methylene-bis-acrylamide	2 g	3 g	8 g	10 g
Acrylamide	38 g	57 g	152 g	190 g
10 × TBE buffer (Section 4.1.9)	125 ml	125 ml	125 ml	125 ml

Make up to 1 litre with double distilled sterile water and filter before use. Store at 4°C.

(vi)   The polymerization time of the gel is important to obtain well formed sample wells. The ideal time is 10 min which will allow time for the gel to be poured and for the sample wells to form satisfactorily.

Polymerization of the gel is initiated by the addition of 1.0 ml of a 10% solution of ammonium persulphate and varying amounts of the catalyst TEMED, 50 μl for a 4% gel, 40 μl for a 6% gel and 25 μl for a 16% gel. The polymerization time will vary with temperature and it is advisable to check the polymerization time of the stock solution before use.

(vii)   Place a support under the sample comb end of the gel plates so that they are at an angle of about 30 degrees to the horizontal.

(viii) Add the ammonium persulphate and TEMED to the acrylamide solution and pour into the well formed by the notch in the gel plate until it is filled. Do not over-fill. The gel solution will migrate down between the two plates with a continuous front. Keep applying the gel solution to the top well otherwise air bubbles will be introduced. Any air bubbles seen forming at the gel front can be prevented by tapping the glass plate at that position. When the gel solution reaches the bottom of the plates remove the support and lay the plates flat. Insert the desired sample comb and allow the gel solution to polymerize.

9.4 Setting up the gel

(i)   Make up 2 litres of 1 × TBE buffer pH 8.3 (Section 4.1.9). Place 1 litre in the lower buffer chamber.

(ii)   Place the bottom of the gel into the lower buffer chamber and secure the top to the upper buffer chamber using G clamps.

(iii)   Fill the upper buffer chamber with 1 litre of 1 × TBE buffer.

(iv)   Remove the sample comb and wash out the wells using a Pasteur pipette to remove any urea and unpolymerized acrylamide.

(v)   Fasten the electrical connections to the thermostatting plate and close the top lid and front safety door.

(vi)   Set the desired temperature on the control box. (Recommended temperature is 65°C.)

(vii)   Pre-electrophorese the gel for 30 min at 2500 V. This will also allow the temperature of the plate to stabilize.

9.5 Electrophoresis of samples

(i)   Denature the samples by heating them at 95°C for 2 min. Cool on ice.

(ii)   Turn off the power supply and open the safety doors. Flush out the wells with buffer using a Pasteur pipette.

(iii)   Using a drawn out capillary load approximately 1 μl of sample into the wells, washing out the capillary with buffer from the lower chamber between each sample.

(iv)   Electrophorese for the desired time and voltage (see *Table 3*).

Table 3. Electrophoresis conditions.

% Acrylamide	Duration (h)	Voltage	Current (mA)	Power (W)	Temperature (°C)
4	14	2500	28	70	65
6	14	2500	22	48	65
16	14	2500	16	40	65

9.6 Autoradiography

(i)   Remove the gel from the electrophoresis apparatus and place on a flat surface.

(ii)   Prise open the gel plates using a thin spatula.

(iii)   The gel will be fixed onto the single gel plate. Place this into a fixing solution of 10% acetic acid and 10% methanol for 20 min, agitating gently at regular intervals. This will remove the urea from the gel and fix the DNA.

(iv)   Remove the gel and wash with distilled water. Remove excess water by carefully laying tissues on the gel. Air dry on the plate at room temperature until the gel no longer feels sticky to the touch.

(v)   In a darkroom lay the gel on a sheet of heavy duty black plastic.

(vi)   Cut a length of XOMAT AR5 X-ray film supplied in a roll by Kodak Ltd to the length of the gel and lay the film on top of the dried gel.

(vii)   Place a gel plate on top of the film and wrap completely with the black plastic.

(viii)   Clamp the sides of the gel 'sandwich' to ensure that the film is held in direct contact with the gel. Place the gel in the dark to autoradiograph for the desired time, as shown in *Table 2*.

Drying the gel onto the glass plates increases the resolution by concentrating the end-labelled DNA fragments from a wet gel thickness of 200 μm to the dried gel thickness of 10−20 μm. Using this method it is not necessary to use screens to enhance the sensitivity. Autoradiographs of three gels with different concentrations of acrylamide containing four samples are shown in *Figure 3*. The samples are loaded in the reaction order G, G+A, T+C, C. It is possible to read at least 500 bases per fragment with sizeable overlaps between each gel. This method was first reported by Ansorge and Barker (3) and has the main advantage in that the gels can be set up and loaded at the same time and electrophoresed for the same time period. However, other methods can be used to achieve long reads on a single gel. Wedge-shaped gels and buffer gradient gels can be used to replace both the 16% and 6% gels (10, and Chapter 2, Section 5.7). Both of these gel systems are easy to pour using ultrathin gels.

10. ADVANTAGES OF MAXAM AND GILBERT SEQUENCING

There are several advantages of sequencing using the chemical cleavage method over the chain-termination method, some of which are detailed below.

(i)   Single-stranded synthetic DNA oligonucleotides can be sequenced directly without any need to clone them. Using a 20% denaturing polyacrylamide gel system and prolonged reaction times the sequence can be determined from the initial 5' base.

(ii)   The sequencing reactions are not affected by long runs of the same nucleotide,

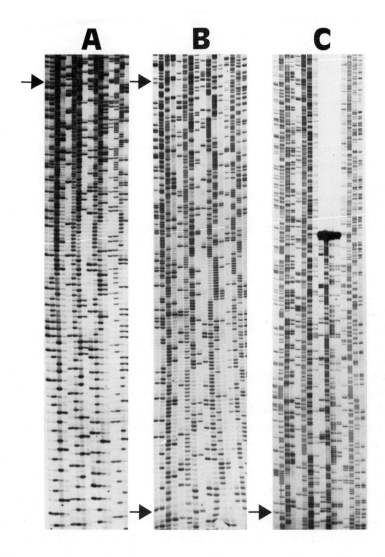

Figure 3. Autoradiographs of three different percentage one metre DNA sequencing gels of differing acrylamide concentrations containing four samples sequenced by the method of Maxam and Gilbert. Each sample is loaded in the order G, G+A, T+C, C. **Gel A** is 16% polyacrylamide and the sequence can be easily read from base 15 to base 150. **Gel B** is 6% and the sequence reads from base 140 to 300. **Gel C** is 4% and the bases read from 300 to 600+. The arrows indicate the crossover positions from the top of one gel to the bottom of the next.

       making it possible to sequence through regions such as poly C, which is often used in cDNA cloning.

(iii)   The procedures of cloning and subcloning are minimized. It is necessary to make only one large plasmid preparation which should provide enough DNA to complete sequencing projects of up to 10 kb. Problems of instability that may be encountered with M13 vectors are circumvented.

(iv)   An example of a specific advantage was reported by Hamilton *et al.* (11) where primer extension techniques were used to determine the genuine 5′ end of a tobacco virus RNA-1. To determine whether the extreme 5′ end of the RNA had been cloned and sequenced, primer extension studies were carried out using a synthetic 17-base oligomer. This was designed to be complementary to a position 14 bases away from the 5′ end of the cDNA clone. Using viral RNA as a template, it was shown that the genuine 5′ end was not represented in the cDNA clone and that approximately 50 bases were missing. Also, more than one extension product was found which subsequently correlated with the fact that the RNA consisted of a population of two molecules (TRV RNA-1 and TRV RNA-2) and that the oligonucleotide was binding to two places on each RNA, thus giving rise to at least four major extension products. This made sequencing by dideoxy chain-termination impossible. However, each one of the extension products could be isolated from a polyacrylamide gel and sequenced by the method of Maxam and Gilbert.

11. RECOMMENDED SUPPLIERS

(i)   Acrylamide: Koch-Light Ltd
(ii)   [γ-^{32}P]adenosine 5′ triphosphate: New England Nuclear
(iii)   Agarose: Koch-Light Ltd
(iv)   Alkaline phosphatase: Boehringer Ltd
(v)   Dimethyl sulphate: BDH
(vi)   Formic acid: Koch-Light Ltd
(vii)   Hydrazine: Sigma
(viii) 3-methacryloxypropyltrimethoxy silane: Koch-Light Ltd
(ix)   *N,N*′-methylene-bis-acrylamide: Koch-Light Ltd
(x)   Phenol: Koch-Light Ltd
(xi)   Piperidine: Koch-Light Ltd
(xii)   T4 polynucleotide kinase: Amersham
(xii)   Repelcote (for siliconizing): BDH
(xiv)   Urea: Koch-Light Ltd
(xv)   Horizontal agarose gel apparatus: Koch-Light Ltd
(xvi)   One metre vertical gel apparatus: Koch-Light Ltd

12. ACKNOWLEDGEMENTS

I wish to thank W.D.O.Hamilton for his contributions to this chapter and also M.G.Jarvis.

13. REFERENCES

1. Maxam,A.M. and Gilbert,W. (1977) *Proc. Natl. Acad. Sci. USA*, **74**, 560.
2. Maxam,A.M. and Gilbert,W. (1980) In *Methods in Enzymology.* Grossman,L. and Moldave,K. (eds), Academic Press, London and New York, Vol. 65, p. 499.
3. Ansorge,W. and Barker,R.F. (1984) *J. Biochem. Biophys. Methods, 9*, 33.
4. Sanger,F. and Coulson,A.R. (1975) *J. Mol. Biol., 94*, 441.
5. Sanger,F., Nicklen,S. and Coulson,A.R. (1977) *Proc. Natl. Acad. Sci. USA, 74*, 5463.
6. Vieira,J. and Messing,J. (1982) *Gene, 19*, 259.

7. Yanisch-Perron,C., Vieira,J. and Messing,J. (1985) *Gene, 33*, 103.
8. Rüther,U., Koenen,M., Otto,K. and Müller-Hill,B. (1981) *Nucleic Acids Res.,* **9**, 4087.
9. Garoff,H. and Ansorge,W. (1981) *Anal. Biochem.,* **115**, 450.
10. Biggin,M.D., Gibson,T.J. and Hong,G.F. (1983) *Proc. Natl. Acad. Sci. USA,* **80**, 3963.
11. Hamilton,W.D.O., Boccara,M., Robinson,D.J. and Baulcombe,D.C. (1987) *J. Gen. Virol.,* **68**, 2563.

CHAPTER 6

RNA sequencing

DAVID A.STAHL, GUIDO KRUPP and ERKO STACKEBRANDT

1. INTRODUCTION

The first complete nucleotide sequence of a gene was published in 1964 by Robert Holley and associates. (1). This sequence was inferred from the transcript of the gene encoding yeast alanine tRNA. Determination of the 77 nucleotide sequence required 1 g of alanine tRNA, recovered from 200 g of bulk tRNA, isolated from 140 kg of baker's yeast and nine years of effort. Holley was awarded the Nobel prize in 1968. This effort compares with the thousand or so nucleotides of sequence now determined from a single high resolution sequencing gel. Recombinant DNA and associated sequencing technology have revolutionized the sequencing of biopolymers and it is now generally far easier to sequence the gene directly than to work with the products of transcription or translation. Yet, from the early sixties to the late seventies the world of nucleic acid sequencing was an RNA world. The history of nucleic acid sequencing technology is largely contained within the history of RNA sequencing. Although the early sequencing technology is cumbersome by contemporary standards, much remains useful in specialized applications. This chapter does not address all of the earlier technology, nor would it be appropriate to do so. Rather, those classical approaches that retain current utility are discussed in detail. Associated but less generally useful techniques are referenced or discussed in passing.

The chapter is organized by historical chronology into four sections. The first describes the use of chromatographic and paper electrophoretic techniques. This will mostly detail variations of two-dimensional fingerprinting techniques for sequence determinations. The second unit describes polyacrylamide gel sequencing of end-labelled RNAs. The third component is the description of primer-extension RNA sequencing reactions using reverse transcriptase which also uses high resolution polyacrylamide sequencing gels, as do most of the protocols described in other chapters of this volume. A final section describes the identification of modified nucleotides and the combined use of gel and chromatographic techniques for rapid sequencing of highly modified RNAs. An overview of the methodology covered in this chapter is shown in *Figure 1*. Detailed descriptions of sequencing gel set-up, electrophoresis and film exposure are covered in protocols within other sections of the book. However, a compilation of tips, tricks and trouble-shooting for the running of high resolution sequencing gels is included in the appropriate sections. These suggestions should complement descriptions in other chapters of this now standard technique.

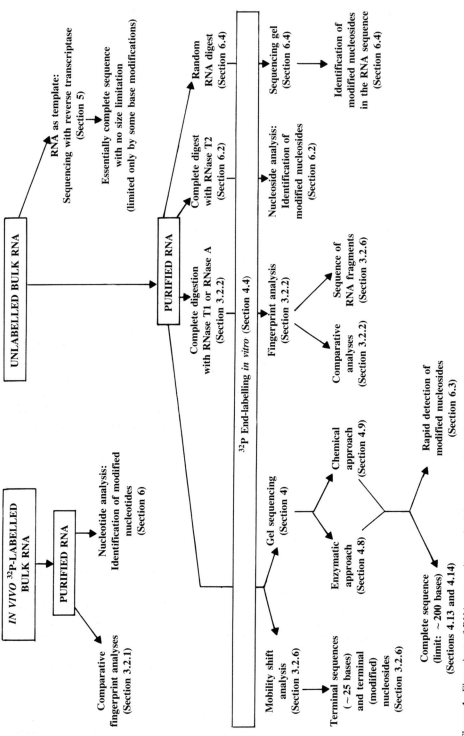

Figure 1. Flow chart of RNA sequencing methodology covered in this chapter.

2. SUMMARY OF THE BASIC TECHNIQUES AND THEIR APPLICATIONS

2.1 Two-dimensional chromatographic techniques

A number of two-dimensional (2-D) systems have been described for the separation of RNA mixtures labelled either *in vivo* or *in vitro*. The two most common use cellulose or gel supports, the selection of which depends on the size and the number of fragments to be separated. The gel supports are most useful for analysis of larger RNAs (10−400 nucleotides), the cellulose supports for smaller (1−25 nucleotides). De Wachter and Fiers (2) have described the methodology and advantages of 2-D gel electrophoretic separation of larger RNAs (10−400 nucleotides) and the reader is referred to this for a detailed description. This chapter will address 2-D high voltage electrophoresis (HVE) on cellulose acetate and DEAE paper (3,4) and a combination of HVE and ascending chromatography on paper (4) or thin layer plates (5−7).

2.1.1 *Chromatographic sequencing of uniformly labelled RNA*

The first two-dimensional separation of RNA labelled *in vivo* (and enzymatically fragmented) used high voltage electrophoresis on cellulose-acetate strips in the first dimension and DEAE cellulose paper (3,4) in the second. The large paper area (45−100 cm) facilitated separation of a larger number of RNA fragments (up to 20 nucleotides in length) than on thin layer plates (5). Individual spots were cut from the paper, RNA digested (either on the paper or following elution) with a variety of enzymes (RNase T1 and T2, RNase U2, pancreatic RNase, spleen acid RNase, spleen exonuclease, snake venom phosphodiesterase, or nuclease P1) and the resulting subfragments separated by one-dimensional electrophoresis on DEAE paper. Non-identifiable fragments were digested a third time, separated one-dimensionally and the sequences reconstructed by 'logical arguments'. Although an improved method (4) has been used in the generation of 16S rRNA 'catalogues' of hundreds of prokaryotes, this approach has few contemporary applications. Therefore only a short outline of the characteristics of a RNase T1 fingerprint will be covered in this context and the reader is directed elsewhere for detailed sequencing protocols (4,8,9).

2.1.2 *Chromatographic sequencing of end-labelled RNA*

The introduction of end labelling marked a major turning point in RNA sequence determination. This made possible the rapid analysis of RNAs from sources where labelling *in vivo* was not feasible, using both chromatographic and polyacrylamide gel sequencing techniques. T4 polynucleotide kinase and γ-labelled nucleoside triphosphates are used to introduce a radioactive phosphate (^{32}P) at available 5'-hydroxyl groups. The 5'-phosphate group of bacterial RNAs must be removed by alkaline phosphatase prior to labelling. Labelling the 5' terminal phosphate of eukaryotic mRNA requires the preliminary removal of the cap structure by either a decapping enzyme or chemical treatment prior to removing the 5'-phosphate group. Since digestion of RNA (e.g. for fingerprinting) by most nucleases generates fragments with 3' terminal phosphates, 3'-phosphatase-free T4 polynucleotide kinase or 3' dephosphorylated RNA should be used to avoid analysis of mixtures of RNA fragments.

Chromatographic sequence analysis of end-labelled RNA requires cleavage to give a family of all possible fragments. Since one end is labelled, only those intermediates with a ^{32}P-labelled terminus are seen following two-dimensional fractionation and autoradiography of the digest. Fragments are generated by enzymatic (nuclease P1 or snake venom phosphodiesterase) or chemical hydrolysis. In order to obtain all intermediates from enzymatic digestion, multiple times of digestion are pooled (time-dependent degradation 'kinetic'). The use of alkaline hydrolysis generates all intermediates in one step and is therefore preferred. Chromatography involves essentially the same procedure as described for the generation of the fingerprint, i.e. by HVE for the first dimension and TLC for the second.

Sequence is inferred by 'mobility shift' or 'wandering spot' analysis (6,10) of the 2-D chromatograph. The 2-D mobility shift analysis of an enzymatically cleaved DNA fragment, including R_f values for representative intermediate fragments, is detailed elsewhere (7). Only the identification of alkaline cleaved intermediates by wandering spot analysis is described in detail (11).

2.1.3 *Applications of RNA fingerprinting*

RNA fingerprinting is no longer used to determine complete nucleotide sequences via overlap reconstructions from partial and complete digestion products. Nonetheless, the technique remains useful in specialized applications that do not require complete sequence information.

(i) *Similarity measures and pattern analysis*. The two-dimensional fractionation is widely used when patterns ('fingerprints') of spots are used to characterize and compare RNAs. In general, relatedness among homologous biopolymers (e.g. of ribosomal and viral RNAs) can be evaluated by comparative sequencing of derived oligonucleotides (12) or by comparison of the two-dimensional pattern of the fingerprints. In combination with two-dimensional separation techniques (mainly RNase T1-resistant oligonucleotides) this has been used to characterize mRNA, viruses, viroids, obligate symbiotes, extremely slow growing bacteria and eukaryotes (13). These measures have been applied to studies of evolution and epidemiology (14,15). More recently, fingerprinting of mixed 5S rRNAs directly isolated from the environment has been used to evaluate the complexity and composition of natural microbial communities (16).

(ii) *Identification of modified nucleotides*. The identification of post transcriptionally modified nucleotides requires the direct analysis of the transcript. Fingerprinting and chromatographic techniques remain the only means for characterization.

(iii) *Gene expression*. Fingerprinting remains useful for certain measures of gene transcription, for example the expression of exogenously introduced 5S rRNA genes in *Xenopus* oocytes (17).

(iv) *Structural probing*. RNA/RNA or RNA/protein cross-linking studies also generally require direct analysis of the cross-linked products. Although the products of structure-dependent chemical and enzymatic modification have been evaluated by fingerprinting, primer extension procedures now provide more powerful and more rapid evaluation of modification (18).

2.2 **Gel sequencing of end-labelled RNAs**

For the sequence analysis of larger RNA molecules, methods based on the use of denaturing polyacrylamide gels are generally used. Here, the end-labelled RNAs (3′ or 5′) are subjected to base-specific enzymatic or chemical cleavage reactions. The products are separated according to size on polyacrylamide gels followed by autoradiography. From such autoradiographs, about 120 nucleotides of sequence can be deduced from a single end-labelled molecule. In general, less sequence can be read from the terminus of an RNA molecule than for a DNA molecule. Presumably, this is the consequence of greater stability of RNA structure relative to that of DNA. Therefore, fragments have to be generated, isolated and sequenced for the complete analysis of larger RNAs. Fragments can be generated by mild nuclease treatments in high salt buffers (19) or preferentially by site-specific cleavage with RNase H and DNA oligonucleotides (20).

2.3 **Primer-extension sequencing of RNA**

The use of RNA templates in primer-extension dideoxynucleotide sequencing reactions was introduced soon after the inception of chain-termination sequencing protocols for DNA (21−23). The technique differs little from reactions using DNA as template except for the requirement for reverse transcriptase. This enzyme will use either DNA or RNA as template and is also suitable for use in sequencing reactions with a DNA template. The enzyme catalyses more uniform incorporation of dideoxynucleotides in the growing transcript than the Klenow fragment of DNA polymerase, resulting in more uniform banding of sequencing gels. These reactions also tend to have lower backgrounds than sequencing gels of Klenow reaction products.

The greatest (single) advantage of sequencing with RNA templates is the elimination of cloning. Disadvantages include the need to know a priming site sequence and the limitation to a single strand sequence determination. There is no straightforward way to obtain complementary strand information for the validation of sequence or for resolution of sequencing gel ambiguities (e.g. band compression or non-uniform dideoxy-nucleotide incorporation). Also, greater care must be taken to isolate intact template. Modified nucleotides can cause the premature truncation of the transcript. Nevertheless, the technique is rapid, given the above provisos, and generally of high fidelity.

2.3.1 *Applications of primer-extension sequencing*

(i) *Message and stable RNA sequencing.* Although applicable to virtually any RNA, the most concerted application of this technique has been to sequencing the transcripts of homologous genes. For such studies the identification of a common priming site or sites among transcripts allows rapid sequence comparisons of adjacent variable regions. This has been used to compare immunoglobulin mRNA sequences, taking advantage of constant region priming sites adjacent to variable regions (24,25). Another common application has been for comparative sequencing of ribosomal RNAs (26,27). As do the immunoglobulin mRNAs, these RNAs also offer variable regions (organism to organism variation) adjacent to conserved regions of sequence.

(ii) *Structural mapping and footprinting.* RNA structure has been analysed by chemical and enzymatic probing followed by primer-extension of the modified product (18). This technique takes advantage of reverse transcriptase pausing or stopping at nucleotides modified by structure-specific chemical probes (e.g. kethoxal or dimethyl sulphate) or running off at sites of nuclease cleavage. Under conditions of limited modification, the relative band intensity of truncated transcripts corresponds to the relative reactivity of the nucleotide. The same approach has been used to determine protein binding sites following chemical modification of RNA/protein complexes (28).

2.4 **Identification of modified nucleosides**

The *in vitro* labelling techniques permit several approaches for the analysis of modified nucleosides.

2.4.1 *Analysis of total nucleoside composition*

RNA is digested completely with RNase T2 and after $5'$ ^{32}P-labelling the products are separated by two-dimensional TLC (6) or by HPLC. The latter gives superior resolution and is quantitative but requires about 10-fold more material (29).

2.4.2 *Idiosyncratic cleavage or mobilities*

The rapid sequencing techniques described can be used for detection and preliminary identification of modified nucleotides. This is based upon variations in cleavage or altered mobilities that can be attributed to known modified residues (30).

2.4.3 *Direct identification and location of modified nucleosides*

An approach developed initially by J.Stanley and S.Vassilenko (31) and subsequently modified (32), is particularly useful for sequencing RNAs rich in modified residues, for example tRNA. In brief, the approach relies upon partial digestion of the RNA such that only a single break is introduced per strand. The resulting fragments are labelled at their $5'$ termini with polynucleotide kinase and $5'$-[γ-^{32}P]ATP and fractionated by size on a high resolution acrylamide gel. The $5'$ terminal nucleotide of each fragment is determined chromatographically following release of the $5'$ nucleotide monophosphate by nuclease digestion. The advantages of this approach include the ability to identify modified nucleotides by two-dimensional chromatography and the resolution (as necessary) of regions of band compression by wandering spot sequence analysis (Section 3.2.2).

3. ELECTROPHORETIC AND CHROMATOGRAPHIC TECHNIQUES

3.1 **Materials**

There is overlap between the enzymes (properties of which are shown in *Figure 2*) and materials used in the various sequencing methodologies. Rather than repeatedly list them under each section, only specialized items and those reagents not previously listed are included in succeeding sections. Autoclave water, buffers and heat-resistant vessels. Work with virgin plastic or baked glassware. Use sterile distilled water for making up all solutions. Working with disposable gloves is recommended. Centri-fugations ('spins') are in a table top Eppendorf microcentrifuge, if not otherwise defined.

Enzyme	Reaction catalysed	Commercial source
Alkaline phosphatase (from calf intestine)	Removal of all terminal phosphate groups	Boehringer Mannheim
Tobacco acid pyrophosphatase	Cleavage of pyrophosphate groups (in cap structures)	Promega
Polynucleotide kinase	Transfer of the γ-phosphate of ATP to the 5'-hydroxyl of nucleic acids (also contains a 3'-phosphatase activity)	New England Biolabs
Polynucleotide kinase (free of 3'-phosphatase)	Mutant enzyme without phosphatase activity	New England Nuclear; Boehringer Mannheim
RNA ligase	Adds pCp (or other donor RNA with 5'-phosphate) to the 3'-hydroxyl of RNA	Pharmacia–LKB; New England Biolabs
Poly A polymerase	Adds ATP (or 3'-deoxy ATP) to the 3'-hydroxyl of RNA	New England Nuclear
RNase H	Endoribonuclease: cleaves RNA only in DNA–RNA hybrids, producing RNA with 5'-phosphate termini	Pharmacia–LKB; Boehringer Mannheim; BRL
RNase A	Endoribonuclease: Cp/N and Up/N	Pharmacia–LKB; Boehringer Mannheim; BRL
RNase T1 Note: Boehringer uses a 100-fold lower unit definition	Endoribonuclease: Gp/N	Pharmacia–LKB; Boehringer Mannheim; BRL
RNase U2	Endoribonuclease: Ap/N (also Gp/N)	Pharmacia–LKB; Boehringer Mannheim; BRL
RNase Cl-3	Endoribonuclease: Cp/N (also Ap/N, Gp/N)	Boehringer Mannheim
Nuclease S7	Endonuclease: Np/A and Np/U	Boehringer Mannheim
RNase Phy M	Endoribonuclease: Np/U, Np/A, Np/G	Pharmacia–LKB; BRL
RNase T2	Unspecific endoribonuclease: Np/N	Calbiochem; BRL
Nuclease P1	Unspecific endonuclease: N/pN	Pharmacia–LKB
Snake venom phosphodiesterase from Crotalus adamanteus	Unspecific 3'-5'-exonuclease: N/pN (requires RNA with free 3'-hydroxyl group)	Pharmacia–LKB
Spleen phosphodiesterase	Unspecific 5'-3'-exonuclease: Np/N (requires RNA with free 5'-hydroxyl group)	Boehringer Mannheim
AMV reverse transcriptase	RNA-dependent DNA polymerase (uses RNA or DNA primers)	Seikagaku America Inc.; Boehringer Mannheim
Terminal deoxynucleotidyl transferase	Adds dNTPs at the 3' end of DNA	Boehringer Mannheim
DNase I (RNase-free)	Unspecific deoxynuclease: N/pN	Promega
Apyrase (grade III)	Hydrolysis of pyrophosphate bonds in ATP	Sigma

Figure 2. Properties and suppliers of nucleic acid modifying enzymes.

3.1.1 *Enzymes and reagents*

(i)   RNase A, alkaline phosphatase from calf intestine (special quality for molecular biology), Boehringer-Mannheim Biochemicals.

(ii)   Polynucleotide kinase, New England Biolabs. Polynucleotide kinase (3'-phosphatase-free), Boehringer-Mannheim Biochemicals or Du Pont NEN Research Products (NEN).

(iii)   RNase T1, RNase U2, Sankyo (Calbiochem) or Boehringer-Mannheim Biochemicals. (Note: Boehringer uses a 100-fold lower unit definition for both enzymes.)

(iv)   RNase T2, Sankyo or Bethesda Research Laboratories.

(v)   Nuclease P1, Pharmacia.

(vi)   Snake (*Crotalus adamanteus*) venom phophodiesterase, Pharmacia.

(vii)   5'-[γ-^{32}P]ATP (7000 Ci mmol^{-1}; 'crude'), ICN Biomedicals Inc.

(viii) 5'-[^{32}P]pCp (3000 Ci mmol^{-1}), Amersham Corporation or NEN.

3.1.2 *Equipment and supplies*

(i)   High voltage electrophoresis tanks (Savant Instruments Inc.) (8).

(ii)   Vacuum centrifuge with a 60-sample rotor (Speed Vac Concentrator, Savant).

(iii)   Bench centrifuge with swing out buckets to hold 60 Eppendorf reaction vials (Beckman TJ6).

(vi)   X-ray film from Fuji (e.g. RX).

(v)   Lightning Plus intensifying screens from DuPont-Cronex.

(vi)   Cellulose acetate strips, Schleicher & Schuell, CA-250/0 (1.5 × 100 cm).

(vii)   Chromatography paper, Whatman DE 81 (45 × 55, and 45 × 100 cm). Cellulose thin layer plates 20 × 20 cm, Merck 5716.

(viii) Radioactive ink: solution of a ^{14}C-labelled compound in normal ink. Use 50 μCi μl^{-1} for exposure times of 1 min to overnight; 2 μCi μl^{1-} for longer exposures.

(ix)   Permanent radioactive markers: fix tiny pieces of filter paper (soaked in radioactive ink) to a strip of transparent Scotch tape and cover with a second strip. Fix with transparent tape at the corners of gels (etc) for accurate alignment of gel and exposed film.

3.1.3 *Autoradiography*

Use pre-flashed films (33) to increase the sensitivity of the film to ^{32}P (approximately half of the otherwise necessary exposure time). Use the pre-flashed film with the sensitized side facing the surface of the gel or paper.

3.2 **Two-dimensional chromatography**

3.2.1 *Two-dimensional chromatography of uniformly labelled RNA*

Uniformly labelled RNA was traditionally isolated from cells grown in low phosphate medium supplemented with ^{32}P (as phosphoric acid). The reader is referred to other references for details of media preparation, cell growth and nucleic acid isolation (34). Conditions for the nuclease digestion (RNase T1 or RNase A) and dephosphorylation of purified RNA are given in *Table 1*. The following description covers only the major

Table 1. RNase digestion.

A. Cleavage

1. Dissolve 0.2 mg of RNase T1 (1000 U) and 150 μg of RNase A in 1 ml of water each.
2. Redissolve 40 μg of RNA in 80 μl of water, add 5 μl of 1 M Tris−HCl (pH 8.0) and either 2.5 units of RNase T1 or 0.4 μg of RNase A.
3. Incubate for 3 h at 37°C
4. If no dephosphorylation is required, dry the oligonucleotides under vacuum by spinning in a Speed Vac Concentrator. Resuspend the pellet in 2 μl of water.

B. Dephosphorylation

1. Add 60 mU (7.5 U ml^{-1}) of alkaline phosphatase, 10 nM ZnCl$_2$ and/or MgCl$_2$ after step 3 and incubate for 3 h at 50°C.
2. Stop the reaction by adding 15 μl of nitrilotriacetic acid (0.04 M, pH 7.2) and incubate for 20 min at 20°C.
3. Boil at 100°C for 2 min.
4. Dry under vacuum using a Speed Vac Concentrator. Resuspend in 2.4 μl of water.

steps of the procedure and the reader is referred to original references (3−5,8,9) for details and modifications.

(i)   First dimension. Soak a cellulose acetate (CA) strip in electrophoresis buffer containing 0.3 M ammonium formate and 7 M urea adjusted to pH 3.5 with a few drops of pyridine. Support a region 5−7 cm from one end of the strip on two (0.7 cm) glass rods that are 2 cm apart. Blot this region with absorbent paper to remove surplus buffer. Apply sample as a small spot in the middle of the strip. Apply dye (1% xylene cyanol, 2% fuchsin, 1% orange C) on each side of the digest. Wait for the sample to be absorbed and remove excess buffer from the rest of the strip by blotting.

(ii)   Place the strip over the electrophoresis rack and submerge it in a HVE tank containing 'white spirit' (Varsol or a comparable solvent) as cooling fluid above the electrode vessels, with the sample near the negative electrode reservoir. Electrophorese at 50−100 V cm^{-1}. Most oligonucleotides migrate between the blue and the yellow markers although some small modified oligomers and uridine-rich fragments may migrate slower and faster than the blue and the yellow dyes, respectively. The length of the CA strip and separation of dyes needed depends upon buffer composition and the number and size of oligonucleotides to be separated (8).

(iii) Second dimension. Remove excess 'white spirit' by blotting with Whatman 3MM paper. Fix the DEAE paper to a glass plate and place the CA strip onto the paper with the blue marker dye at the right side of the short side. Place a pad of five water-soaked Whatman No. 3MM paper strips (1 × 50 cm) on top of the CA strip and transfer the oligonucleotides to the DEAE paper by pressing the paper strips together evenly with the aid of a glass plate.

(iv)   Remove the strip after the transfer (about 2 min) and position the paper on the electrophoresis rack. Wash off the urea by rinsing (with a squirt bottle) the wetted area on the DEAE paper with 70% ethanol and allow to dry. Wet the paper evenly with the second-dimension buffer by squirting liquid from a soft plastic bottle. Common second-dimension buffers include 6.5% formic acid, 0.1 M pyridinium formate, pH 2.3, or 2.5% formic acid, 8.7% acetic acid (v/v), pH 1.9. High salt buffers (3,4) provide better resolution in most regions of the fingerprint, eliminating the need to dephosphorylate oligonucleotides prior to electrophoresis (4). Use 30 V cm^{-1} and 20 V cm^{-1} for separation on short and long paper, respectively. Depending on paper size, buffer and oligonucleotide composition, vary the time of electrophoresis between 16 and 24 h.

(v)   Dry and mark the paper with radioactive ink. Cut the DEAE paper sheets into two or three pieces to match the size of film cassettes (X-ray film) for autoradiography. Expose for $2-12$ h, depending on the activity of the fingerprints.

Oligonucleotides separated on paper migrate within a series of wedge-shaped 'isopleths' (3). Their shapes and positions vary with the second-dimensional solvent. Each isopleth contains oligomers characterized by the same number of U residues. As schematically depicted in *Figure 3* for T1 digested RNA, the fastest moving isopleth contains oligomers with a high number of uridine residues, followed by sets of oligomers in which the uridine content is decreased by one per isopleth. Within each isopleth oligomers having the same number of bases are arranged along almost vertical lines (isomeric lines). On a given isomeric line oligomers are separated by their relative A (versus C) content. The oligomers with greatest A content migrate slowest in the second dimension. Oligonucleotides with the same gross composition but different sequences occupy the same, or nearly the same position in the map (for exceptions see reference 4).

Since the position of an oligomer is determined largely by nucleotide composition, its gross composition can be inferred from neighbouring oligonucleotides of known sequence or known composition. The composition or sequence of reference oligonucleotides are determined by position as shown in *Figures 3* and *5*. Likewise, the position of oligomers of known sequence is predictable. The classical chromatographic techniques to deduce sequence by 'logical arguments' have been replaced by end-label analysis and will not be covered here (see refs 4 and 8 for detailed descriptions). Note however that the two-dimensional fractionation described here also has applications for the analysis of end-labelled RNA (Section 3.2.2).

3.2.2 *Two-dimensional chromatography of end-labelled RNA*

The ease of labelling and analysis has made this the preferred approach to fingerprinting and sequence determination. The first-dimensional separation of end-labelled RNA on cellulose-acetate strips resembles the conditions used for uniformly labelled RNA. Separation in the second dimension is performed on DEAE cellulose thin layer plates rather than the traditional paper. The thin layer plates offer a much smaller separation area and consequently, either a small number of fragments are completely resolved or conditions are applied which favour the separation of fragments of a defined size and composition. Fractionation in the second dimension is by homochromatography or ammonium-formate gradient chromatography. During homochromatography, radioactive nucleotides are displaced by a series of unlabelled oligonucleotides of different valency or affinity ('Homomixture') by saturation of DEAE groups. The smaller ones (with lower binding capacity) are replaced by larger ones and therefore move faster.

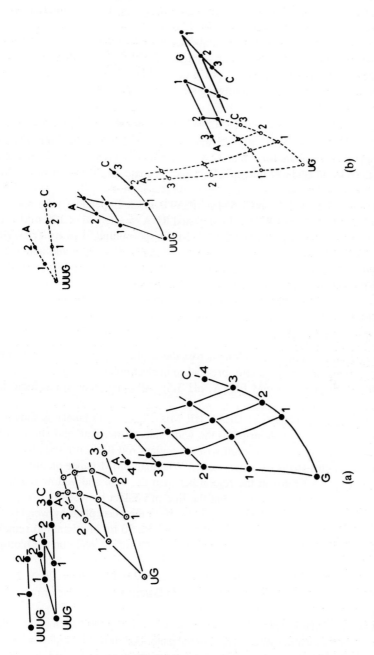

Figure 3. Schematic correlation between the gross nucleotide composition of a RNase T1 oligonucleotide and its position in the two-dimensional system using 7% formic acid for the second dimension on DEAE-cellulose paper (8). Note the influence of the uridine content on the formation of 'isopleths'; (**a**) nucleotides carrying and (**b**) nucleotides lacking a 3'-phosphate residue. With permission of G.G.Brownlee and North-Holland Publishing Co.

The disadvantage of this method is the relatively high content of unlabelled RNA ($\sim 500-1000$ μg cm^{-2}) that may cause problems in subsequent sequence analysis. Currently, ammonium-formate gradient chromatography is used as an alternative for the 'Homomixtures' in the second dimension (7,11).

The protocol for nuclease digestion of unlabelled RNA is given in *Table 1*, that for alkali cleavage in Section 3.2.5. Note that undigested end-labelled RNA can also be analysed by two-dimensional separation for the determination of about 25 nucleotides of sequence from either end. Use the following steps for labelling the RNA. See Section 4.6 for troubleshooting.

(i) Dephosphorylation. Dissolve $0.1-0.2$ A_{260} units of RNA in 10 μl of water, add 0.5 μl of 1 M Tris$-$HCl, pH 7.5 and 0.5 μl of calf intestine phosphatase (20 U per ml). Incubate for 1 h at 37°C. Terminate the reaction by adding 2 μl of nitrilotriacetic acid (40 mM, pH 7.2) and incubate for 20 min at 37°C. Boil for 1 min.

(ii) 5'-Labelling. Dry $0.1-1$ mCi 5'-[γ-^{32}P]ATP in an Eppendorf tube and add 10 μl of water, 2.5 μg of RNase T1-digested RNA (5 μl), 1 μl of 100 mM DTT (dithiothreitol), 1 μl of 0.2 M MgCl$_2$/32 mM spermidine, 1 μl of Tris$-$HCl buffer (0.25 M, pH 7.5), and 1 μl of 50 mM ATP. Mix, spin and add 1.2 μl of polynucleotide kinase, 3'-phosphatase-free (4 U). Incubate for 45 min at 37°C and terminate by boiling for 2 min. Dry the pellet.

(iii) 3'-Labelling. Follow the protocols given below in *Tables 10* and *11* for the preparation of [5'-^{32}P]pCp and ligation reaction.

Continue as follows for generation of the fingerprint.

(iv) Apply sample in a line 1.2 cm long across the width of a cellulose acetate strip (1.5 $\times$ 100 cm or 1.5 $\times$ 55 cm), pre-soaked in electrophoresis buffer (5% acetic acid, 0.05% pyridine, 5 mM EDTA, pH 3.5). All other steps are as indicated under Section 3.2.1.
cellulose thin layer plate (Cel DEAE HR-Mix-20, 0.2 mm layer) as shown in *Figure 4*. Apply long CA strips to the wide side of the plate and short strips to the narrow. Lay the strip lengthwise on a glass rod (0.4 cm) and place five 45 cm (or 25 cm) strips of water-soaked Whatman 3MM paper on either side overlapping it ($\sim$3 mm) in the region between the blue and yellow markers. Place the plate onto the strip so that the line of contact between the strip and plate runs 2 cm from the edge of the plate. Hold the glass plate by weights for 15 min. Rinse the transfer region of the plate with 50 ml of ethanol to remove HVE buffer salts. Apply marker dye near the edges of the plate on the blotting line.

(vi) Place a filter paper wick (three sheets of Whatman 3MM filter paper, 20 $\times$ 20 cm or 20 $\times$ 40 cm) on top of the DEAE plate and fasten them with one or two metal binder clip(s).

(vii) Use two chromatography chambers placed on top of each other for chromatography over the 40 cm length of the plate. For chromatography on 20 cm plates use a glass plate covered chromatography tank (24 $\times$ 44 $\times$ 28 cm) containing two glass troughs (42 $\times$ 8 $\times$ 4 cm), one for pre-chromatography, the second for chromatography. Support the plates on notched plastic rods placed between two ordinary glass plates of the same dimensions.

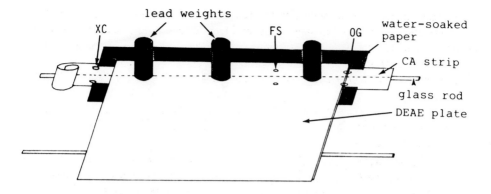

Figure 4. Schematic illustration of transfer of oligonucleotides from cellulose-acetate high voltage electrophoresis strip to DEAE-cellulose TLC plate. XC, xylene cyanol; FS, acid fuchsin; OG, orange G.

Table 2. Preparation of 'Homomixture'.

1.	Dissolve 3% yeast sodium nucleinate in 7 M urea and add 20, 15, 10 and 2 ml of 10 M KOH to 2 l to obtain 100 mM, 75 mM, 50 mM and 10 mM 'Homomixtures'.
2.	Hydrolyse for 20 h at 65°C and cool slowly for 2 h at room temperature.
3.	Adjust pH to 4.7 with glacial acetic acid.

(viii) Pre-chromatograph at 70°C by allowing water to migrate for 15 cm. Replace water with the second-dimension buffer, for example 'Homomixture' (*Table 2*) or a linear ammonium formate gradient (0.30−0.40 M in 7 M urea, 1 mM EDTA).

(ix) Terminate chromatography when the blue marker dye has almost reached the upper edge of the plate (~3 h for small fragments) or when the orange G marker has migrated at least 18 cm (~6 h for fragments >6 nucleotides); smaller oligomers have by then migrated into the wick.

(x) Remove the plate from the tank, remove the wick, let the plate dry and mark it with radioactive ink. Locate the position of oligomers by autoradiography.

3.2.3 *Interpretation of the two-dimensional chromatograph of end-labelled RNA*

Replacement of HVE by chromatography changes the separation behaviour of oligonucleotides considerably (*Figure 5*). Oligomers of equal length but different composition are separated in the form of a triangle (for RNase T1-cleaved RNA) or in the form of a straight line (for RNase A-cleaved RNA).

For RNAse T1 digestion products the corners of each triangle are occupied by fragments consisting of one kind of oligonucleotide ending in G (*Figure 5*A). C_nG and U_nG occupy the left and right corners of the long side of the triangle, respectively. A_nG runs faster than C_nG in the first but markedly slower in the second dimension. Oligomers of a given size with the same uridine content are arranged in diagonal lines, with cytidine-rich fragments migrating slower in the first but faster than adenosine-rich oligomers in the second dimension. Fragments with the same size and gross composition in general occupy the same position in the fingerprint although deviations

149

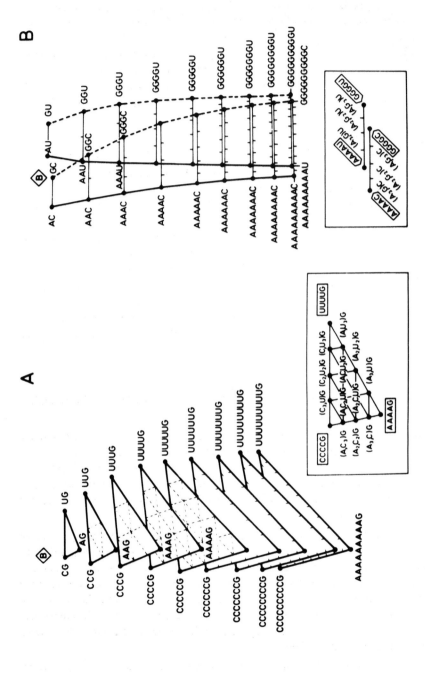

Figure 5. Schematic correlation between the gross nucleotide composition of a RNase T1 (**A**) and a RNase A (**B**) oligonucleotide and its position in homochromatography or ammonium formate gradient chromatography on DEAE-cellulose thin layer plates (35). Note the difference in separation behaviour as compared to that of *Figure 3*. With permission of the authors and IRL Press.

are known (11). Triangles of oligonucleotides are not separated completely from each other (*Figure 5A*), but overlap. The degree to which overlapping occurs depends to some extent on the molarity of the second-dimension buffer. As seen in *Figure 5A*, the distinction between oligomers of two size classes, i.e. the boundaries between two triangles, is not immediately apparent from the fingerprint.

It is more difficult to correlate position with composition for RNase A fingerprints. Not only is the number of oligomers likely to be increased but the horizontal lines are barely resolved because the 3' terminal pyrimidines have only a minor influence on the separation of oligomers of a given size in the second dimension (*Figure 5B*).

3.2.4 *Extraction of RNA from cellulose*

Use the following procedure for recovering oligonucleotides from the thin layer plates.

(i)   Fix developed film underneath the TLC plate. Illuminate from below and align by means of the radioactive marker dots.

(ii) Prepare drawn-out glass tubes (4 × 0.3 cm), stopped with a small piece of sterile cotton wool.

(iii) Scrape and suck the RNA containing DEAE cellulose from the spot of choice into the glass tube using a vacuum line attached to a water aspirator.

(iv) Place the tube through a punched lid of an Eppendorf tube.

(v)   Elute RNA from the cellulose and into the vial by three washes with 80 μl of 1 M NaCl using a swinging-bucket rotor (750 g, 2 min).

(vi) Precipitate RNA with 3 vol of ethanol at $-20°C$ for at least 2 h, centrifuge at 1000 g for 20 min, discard the supernatant and dry the pellet in a Speed Vac Concentrator.

3.2.5 *Hydrolysis of eluted RNA*

Digest the end-labelled oligomers enzymatically or by base hydrolysis for mobility-shift sequence analysis.

(i) *Nuclease P1 digestion.* Dissolve $1-5$ A_{260} units of carrier and labelled RNA ('Homomixture' plates already contain about $0.5-1$ mg RNA cm^{-2}) and 50 ng of nuclease P1 per A_{260} unit in 10 μl of 50 mM ammonium acetate, pH 5.3. Incubate at 20°C. Pipette the aliquots after 0 (1 μl), 5, 10 and 20 min (3 μl each) onto 1 μl of Na$_2$EDTA. Hold on a dry-ice/ethanol bath until the reaction set is complete. Boil for 4 min and dry under vacuum.

(ii) *Snake venom phosphodiesterase.* Dissolve RNA in 19.2 μl of water, 3.3 μl of enzyme (3 mg ml^{-1}) and 2.5 μl of buffer (0.5 M Tris$-$HCl, 0.05 M K$_3$PO$_4$, pH 8.0). Incubate at 20°C for 5 min and transfer to 37°C. Pipette aliquots after 0 min (1 μl), 1, 2, 5, 10, 20, 40, 60, 80 and 120 min (2.5 μl each) onto 1 μl of Na$_2$EDTA and hold in a dry-ice/ethanol bath. Boil for 4 min and dry under vacuum. Note that since high amounts of enzyme are necessary for digestion of RNA isolated from 'Homomixture' plates, it is recommended to use this enzyme only for carrier-free RNA isolated from ammonium formate gradient plates.

(iii) *Chemical hydrolysis.* The following conditions have been developed for carrier-free RNA: dissolve RNA in 10 μl of 50 mM NaHCO$_3$, adjusted to pH 9.2 with 50 mM NaOH (35), and boil for 18 min. Chill on ice immediately, add 1.3 μl of 2.5 M HCl and dry under vacuum.

3.2.6 *Two-dimensional separation: mobility shift analysis*

This technique involves basically the same procedures as described for the generation of the fingerprint by HVE and TLC (Sections 3.2.1 and 3.2.2). A procedure for the simultaneous separation of about 60 samples has been described (11). Note that 5'- and 3'-labelled alkali-cleaved intermediates differ in their migration behaviour. While 5'-labelled fragments carry a 5' terminal phosphate group (pNp, pNpNp, pNpNpNp and so on) 3'-labelled fragments are of the NpCp, NpNpCp, NpNpNpCp and so on series. The lack of a phosphate group makes the 3'-labelled oligomers run slower than the xylene cyanol marker in the first but slightly faster in the second dimension. For 5'-labelled fragments this separation behaviour is reversed. Consider this for the transfer of material to, and the separation of oligomers in, the second dimension.

(i)   Dissolve hydrolysed sample in 2.4 μl of distilled water and apply up to three 1-μl samples to one cellulose acetate strip (1.5 × 55 cm), 6, 18 and 30 cm from the end. If working with a high number of samples, i.e. 12 strips, drying of the pre-soaked strips (Sections 3.2.1 and 3.2.2) must be avoided. Allow the xylene cyanol marker to migrate 6 cm at 5000 V.

(ii) Transfer the material by blotting onto a small (for one sample) or long (for two or three samples) side of a 20 × 40 cm DEAE cellulose thin layer foil (fixed to a glass plate of the same dimensions). Rinse the wetted area with ethanol, fasten a wick and apply marker dyes as described for the primary fingerprint (Section 3.2.2).

(iii) Pre-chromatograph at 70°C and develop at the same temperature by homo-chromatography or ammonium formate gradient chromatography. Depending on the number of plates and the size of oligonucleotides, use either a tank (for several broadside plates) or two chromatography chambers placed on top of each other (for upright plates) as indicated in Section 3.2.2.

For homochomatography, the size of the oligonucleotides to be separated determines the composition of the 'Homomixtures'. Satisfactory separation of oligomers up to 24 bases has been achieved by a mixture consisting of 50:75:100 mM KOH at 1:2:1 (*Table 2*) containing 1 M boric acid.

For ammonium formate chromatography, replace water of the pre-run with 0.28 M ammonium formate, 1 M boric acid, 7 M urea, 1 mM EDTA, pH 4.5. Immediately begin addition (by pumping) and mixing of the more concentrated solution (linear gradient of 0.5−2.0 M ammonium formate, 1 M boric acid, 7 M urea, 1 M EDTA, pH 4.5) to generate a non-linear ammonium formate gradient in the tank (0.28−0.43 M).

(iv) Remove TLC foils from the tank when the xylene cyanol marker has reached the wick. Remove the wicks, separate the foils from the glass plates and dry them. Mark and expose film to the foils (Section 3.1.3).

3.2.7 *Interpretation of the mobility shift patterns*

Separation of intermediary fragments by HVE and TLC generally follows the same rules as discussed for the primary fingerprint (Section 3.2.3). Knowing the influence the addition of a mononucleotide to a mono- or oligonucleotide has on the migration pattern, the cleavage series (sequence) of an oligonucleotide can be directly read as an angular mobility shift between successively longer fragments (6,11) (*Figures 6* and *7*). The addition of Cp and Ap to [^{32}P]Cp, [^{32}P]Ap, [^{32}P]Up or [^{32}P]Gp or to intermediates ending in Cp, Ap, Up or Gp causes a distinct shift to the left, while the addition of Up and Gp causes a shift to the right. The respective effects are more pronounced with the pyrimidine-phosphates than with the purine-phosphates. In the second dimension, intermediates ending in Ap and Gp cause fragments to run slower than oligomers with 3' Cp or Up. In longer sequences containing Ap and Cp only, both nucleotides can be easily distinguished. Long stretches of cytidines almost turn vertically, while adenine stretches tend to give a shift to the right. In any case, the distances between two intermediates decrease with increased length of the oligonucleotide and it is always the direct comparison of the angular shift of the neighbouring intermediates that must be considered in sequence determination.

3.2.8 *Tertiary analysis*

While most sequences can be determined by wandering spot analysis alone, there are cases in which additional (tertiary) analysis is required, e.g. for determination of the 5' terminus or for characterizing co-migrating oligonucleotides.

(i)   Analysis of the 5' terminus is necessary when its identity is not obvious, i.e. when an oligomer has a [^{32}P]Cp or [^{32}P]Ap terminus, or when two or more oligonucleotides co-migrate in the fingerprint (having different 5' termini). In the latter case determination of the 5' terminus of selected intermediate spots from the secondary analysis may help trace the migration pattern of the individual oligonucleotides. Since nuclease P1 binds rather inefficiently to mononucleotides, the fragment eluted for this treatment should be at least a dinucleotide. Otherwise a double-spot will result from 3'-phosphorylated and -dephosphorylated nucleotides (11). The protocol is given in *Table 3*.

(ii)   Analysis of the 3' terminal region of oligonucleotides greater than decamers may be necessary if the xylene cyanol marker has moved 20 cm (broadside foils) or less. Repeat the chromatography of these samples using an extended second dimension in 'Homomixture' (yellow marker dye migrating 18 cm).

(iii)   Use one-dimensional chromatography to discriminate unambiguously between adenosine and cytidine residues within a uridine-rich sequence (*Table 4*).

(iv)   Use 2-D analysis (HVE and TLC) of individual alkaline hydrolysis fragments for further characterization of co-migrating oligonucleotides (*Figures 6* and *7*). One to three week exposures with intensifying screens are generally required because of low activity.

(v)   Post-transcriptional modification of the base generally does not markedly influence chromatographic behaviour (see also Section 6.3). Ribose-2'-O-methyl modifi-

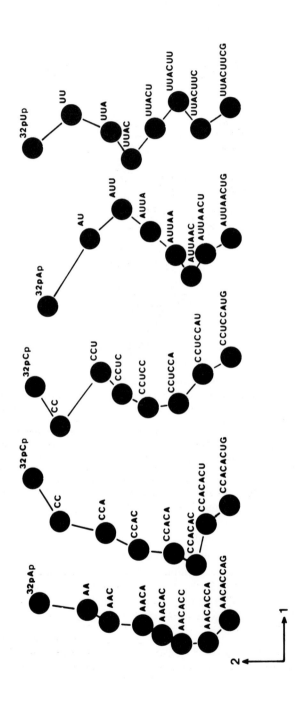

Figure 6. Schematic mobility shift separation pattern. Sequence is inferred from characteristic angular mobility shifts between each oligonucleotide and the nucleotide addition to it (11). With permission of the authors and Academic Press, Orlando.

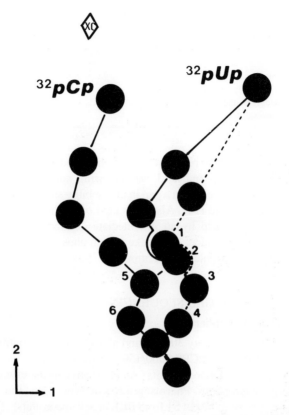

Figure 7. Schematic of sequence determination of comigrating oligonucleotides using tertiary analysis (11). The following sequences could be present: CCCUUAUG, CCUUCAUG, CCUUUACG, CCUCUAUG, UCCCUAUG, UCCUCAUG, UCCUUACG and UAAUACG. The identities of the 5′ termini are obvious from their position relative to the xylene cyanol marker dye position. Sequence determination of isolated intermediates gave the following results: 1, UAA and traces of UCCU; 2, UCCU and traces of UAA; 3, UAAU; 4, UAAUA; 5, CCCUU and UCCUC; and 6, CCCUUA and UCCUCA. The inferred sequences are UAAUACG, CCCUUAUG and UCCUCAUG. Note that the original material isolated from the primary fingerprint contained a mixture of one heptamer with two uridine and three adenine residues and two octamers with three uridine and only one adenine residue. This resulted from the overlap between triangles formed by oligonucleotides which differ in size by one base (see *Figure 5*). With permission of the authors and Academic Press, Orlando.

Table 3. Determination of the 5′ terminus.

1.	Scrape the respective material from the thin layer foil and isolate RNA as indicated in Section 3.2.4.
2.	Digest RNA completely with 4 μl of nuclease P1 (25 ng ml^{-1}) in 0.05 M ammonium acetate, pH 4.5 at 37°C for 3 h or 14 h when digesting material from 'Homomixture' plates.
3.	Apply up to 18 samples to a cellulose thin layer plate (20 × 20 cm) and develop in one dimension in tertiary butanol:H$_2$O:conc. HCl (14:3:3) (6) until the solvent front almost reaches the top of the plate. Migration is as follows: uridine ≫ cytidine > adenine > guanine. Autoradiograph for one to several days. Include markers if internal markers are not present in the digest.

Table 4. Separation of oligonucleotides in one dimension.

1.	Add 100 μl of water to samples remaining after the secondary analysis and precipitate RNA with 300 μl of ethanol for 14 h at −14°C. This step removes salts remaining after alkaline hydrolysis.
2.	Pellet the RNA by centrifugation, dry the samples and redissolve in 3 μl of water.
3.	Apply samples 1 cm apart along a line 2 cm from the bottom edge of a DEAE-cellulose thin layer plate (20 × 30 cm). Fasten a wick, wash the wetted area with ethanol and apply dye mix.
4.	Pre-chromatograph and 'homochromatograph' at 70°C as described in Section 3.2.2. Expose film for one to several days.

cations are detected by resistance of the phosphodiester bond (between a ribose with a methyl group and its 3′ neighbouring nucleotide) to base hydrolysis. However, since this results in a gap in the mobility shift pattern the exact sequence cannot be determined (11). Also, this kind of modification may be overlooked if the gap is occupied by an intermediate derived from a non-modified co-migrating oligonucleotide. Use snake venom phosphodiesterase and nuclease P1 for unambiguous sequence determination (Section 3.2.5).

4. POLYACRYLAMIDE GEL SEQUENCING OF END-LABELLED RNAS

4.1 Equipment and suppliers

(i)   Power supplies. Capacity 3000 V and about 150 mA. Controlled power output is advantageous (e.g. ECPS 3000/150 or Macrodrive 5 from Pharmacia/LKB).

(ii)   Vertical gel electrophoresis units. An alternative to the efficient and expensive units (e.g. Macrophore, Pharmacia/LKB or Sequi-Gen, Bio-Rad) is one of the simple systems (e.g. model S1 from BRL or self-made units). The 'smiling effect' of the gels can be minimized by tightly clamping an aluminium plate to the gel.

(iii)   Biotrap chamber. Schleicher & Schuell (for electroelution of large RNAs).

(iv)   Glass plates. Use a local supplier to cut plate glass to 20 × 40 × 0.5 cm or appropriate size for sequencing unit.

(v)   Spacers and combs. Commercially available or cut from Plastikard (0.35 mm thick, Slater's Ltd, Matlock, Bath, Derbyshire, UK) or Shimstock (e.g. 0.27 mm thick, Artus, PO Box 511, Englewood, New Jersey 07631, USA).

(vi)   Polyethyleneimine impregnated 0.1 mm cellulose thin-layer plates (PEI-cellulose), Brinkman Instruments, Inc.

4.2 Reagents and chemicals

(i)   Acrylamide, *N,N'*-methylene bisacrylamide (2 × crystallized or deionized with Amberlite MB-1 and filtered through Whatman 3MM paper), *N,N,N',N'*-tetramethylethylenediamine, Amberlite MB-1, xylene cyanol FF ('XC') and bromophenol blue ('BB'), Serva.

(ii)   Glycogen (special quality for molecular biology), tRNA from bakers' yeast (carrier RNA) and nucleotides, Boehringer-Mannheim Biochemicals.

(iii)   Dimethylsulphoxide (DMSO), nitrilotriacetic acid (NTA; free acid), diethyl pyrocarbonate, hydrazine (available in 2-ml ampoules), hydroxylamine hydrochloride, diethylamine and aniline, Sigma Chemical Company. Note anhydrous hydrazine is unstable. Purchase small amounts and discard when sequencing reactions are unsatisfactory. To discard old supply safely, mix small aliquots with 3 M $FeCl_3$.

(iv)   Dimethylsulphate (DMS) and sodium borohydride, Merck.

(v) $5'-[\alpha-^{32}P]GTP$ (3000 Ci mmol^{-1}), $5'-[\alpha-^{32}P]ATP$ (3000 Ci mmol^{-1}), Amersham or New England Nuclear.

(vi) $5'-[\alpha-^{32}P]3'$-deoxyadenosine 5'-triphosphate (Cordycepin 5'-triphosphate, 5000 Ci mmol^{-1}), New England Nuclear.

4.3 Enzymes

(i)   Tobacco acid pyrophosphatase and guanylyl transferase, Bethesda Research Laboratories.

(ii)   RNA ligase, Pharmacia.

(iii)   Poly A polymerase, New England Nuclear.

(iv)   RNase Cl 3 and Nuclease S7, Boehringer-Mannheim Biochemicals.

(v)   Nuclease Phy M, Pharmacia or Bethesda Research Laboratories.

4.4 End-labelling of RNA at unique termini

Direct sequence analysis is limited to about 200 nucleotides from a uniquely labelled terminus. If the RNA is small, or if sequencing from both termini will complete analysis, proceed to Section 4.4.2. If internal regions are out of reach (greater than about 200 nucleotides from either terminus), unique internal cuts can be generated by partial digestion with any of a variety of non-specific ribonucleases (*Figure 8*) or by site-specific RNA cleavage by RNase H.

4.4.1 Site-specific cleavage of RNA with RNase H

This method is based upon recognition of DNA−RNA hybrids and cleavage of the RNA moiety by RNase H. The ideal situation is shown in *Figure 9*, and the protocol in *Table 5*. This approach has been used for RNA sequencing (38) and for the specific removal of functionally important RNA segments (39). However, the following limitations have been observed (40).

(i)   The DNA−RNA hybrids can form only with single-stranded RNA. Since RNA helices are generally more stable structures than the hybrid helices, internal structure competes with hybrid formation. On the other hand, this property can be exploited for the analysis of RNA structure (41).

(ii)   Only tetramer hybrids are cleaved specifically. These however may not be stable. Longer oligonucleotides favour hybrid formation but cleavages are less specific; several staggered cleavages may occur at the end of the hybrid or internally. Mixed DNA−RNA−oligonucleotides have been used to alleviate this effect (20).

4.4.2 End-labelling of RNAs

See *Tables 6−12*. All protocols are suitable for 1 to approximately 100 pmol of RNA. In order to reduce the risk of sample losses during ethanol precipitations, include 5 μg of glycogen as carrier.

(i) *Methods for 5'-^{32}P-labelling of RNA*. The different methods for 5'-labelling also

NUCLEOSIDE

	G	m1G	m2G	m7G	m2_2G	A	I	m1A	m2A	m6A	i6A	t6A	ms2_6A	U	D	T	F	s4U	o5U	mam5s2U	C	m5C	ac4C	yW	Q
Enzymatic digests (1)																									
T1	++	(+)	+	−	−	++	−	−	−	−	−	−	−	−	−	−	−	−	−	−	−	−	−	?	?
U2	(+)	−	−	−	−	++	−	−	−	+	−	−	−	−	−	−	−	−	−	−	−	−	−	?	?
H+(a)	large	large	large	large	large	large	normal	large	small	large	large	large	normal	normal	small	large	large	large	small	normal	small	huge	small	huge	huge
S7	−	(+)	−	+	?	−	+	−	−	(+)	++	+	+	++	+	(+)	(+)	+	−	−	−	++	++	+	?
Cl-3	−	?	?	?	?	(+)	?	?	?	(+)	−	−	?	−	(+)	+	(+)	?	+	(+)	−	++	++	?	?
Chemical cleavages(2)																									
G	++	++	+	++	+++	−	+	++	−	++	(+)	?	?	−	−	−	−	−	−	−	−	++	++	+	+
A	+	+	+	+	−	++	++	−	+	(+)	++	?	?	−	−	(+)	−	+	−	−	−	++	++	+	+
U	+	+	−	−	−	−	−	+	−	−	−	?	?	++	+	−	+	?	+	++	−	++	++	+	+
C	−	?	−	?	−	−	−	−	−	−	?	?	?	−	?	−	−	−	+	++	−	++	++	?	+
K(b)	−	−	−	++	−	−	−	−	−	−	?	?	?	−	−	−	−	−	−	−	−	+	+	?	+
Mobility shift analysis(c)	G	G	G	G reverse?	G	A	A reverse?	C reverse?	A	large small				U	U	U	U doublet	U doublet	U	U doublet	C	C	C	?	?

Figure 8. Reaction patterns of modified nucleosides. Cleavage intensities: −, no cleavage; (+), weak; +, moderate; ++, strong. The intensity can vary, however, with the local secondary structure. (**a**) Here the various spacings in the sequence ladder are given; (**b**) the nucleoside is cleaved by aniline without prior modification; (**c**) see also Silberklang *et al.* (6). This scheme was published previously in Nucleic Acids Research (36). Note: 2'-O-methylated nucleosides produce a characteristic gap in the sequencing ladder (acid cleavage) and they are resistant to all enzymes mentioned here (under sequencing conditions). However, they behave like their unmodified counterparts in the chemical reactions. No systematic studies are available to include reverse transcriptase sequencing (Section 5) in this table. As a general rule, however, it was noted that the transcriptase reaction is blocked by modifications which interfere with the formation of Watson Crick base pairs, e.g. like m1A and m3C, but no block occurs with m7G (37).

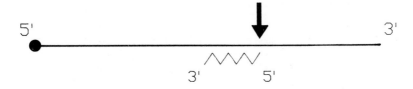

Figure 9. Site-specific cleavage by RNase H. A short DNA oligonucleotide (zigzag line) is hybridized to single-stranded RNA. The RNA is cleaved at the 3′ end of the hybrid (arrow).

Table 5. Site-directed cleavage with RNase H.

1.	Mix approximately 2 pmol of RNA (e.g. 1 μg of 16S rRNA) and 100 pmol of oligonucleotide (e.g. 0.2 μg of hexamer) in 10 μl of 40 mM Tris−HCl (pH 8.0), 4 mM MgCl$_2$, 1 mM DTT.
2.	Heat to at least 50°C (or boil) and slowly cool to 30°C.
3.	Digest the hybrids with 1 unit of RNase H for 30 min at 30°C.
4.	Label the RNA fragments (5′-phosphorylated) with RNA ligase (*Table 10*) or with polynucleotide kinase following dephosphorylation (*Table 6*).

Table 6. Dephosphorylation and 5′-labelling (42).

A. Dephosphorylation

1.	Mix 7 μl of RNA (~1−100 pmol in H$_2$O, e.g. 0.03−3 μg tRNA) with 1 μl of 1 M Tris−HCl (pH 8).
2.	Heat for 2 min at 100°C, chill in ice and spin briefly (denaturation step).
3.	Add 1 μl of alkaline phosphatase (20 mU) and incubate for 1 h at 50°C and spin briefly.
4.	Add 3 μl of 50 mM NTA (nitrilotriacetic acid; pH 7.2) and incubate for 20 min at 50°C to inactivate enzyme.

Use immediately for 5′-labelling or store at −20°C.

B. 5′-^{32}P end-labelling

1.	Denature the dephosphorylated RNA (12 μl) for 2 min at 100°C, chill on ice and spin briefly.
2.	Add 1 μl of 0.2 M MgCl$_2$/32 mM spermidine, 1 μl of 100 mM DTT, 1 μl (~100−200 μCi) of 5′-[γ-^{32}P]ATP, and 1 μl of polynucleotide kinase (4 U). Incubate for 30 min at 37°C.
3.	Add 16 μl of 4 M ammonium acetate (pH 7) and 80 μl of ethanol to terminate the reaction. Freeze (5 min in dry ice or overnight at −20°C) and spin for 10 min. Carefully remove and discard the supernatant. Most of the unused 5′-[γ-^{32}P]ATP remains in the supernatant following ethanol precipitation from high salt.

Use for RNA with 5′ terminal hydroxyl, mono-, di- or triphosphate groups, i.e. for primary transcripts (not capped; see *Table 7*) or processed RNAs. Alkaline phosphatase is used to remove 5′ terminal phosphates followed by phosphorylation of the 5′-hydroxyl group with polynucleotide kinase and 5′-[γ-^{32}P]ATP .

permit the analysis of the structure of the 5′ end. The characteristic structural requirements for the RNA terminus are mentioned for each method.

(ii) *Methods for 3′-labelling of RNAs.* All methods described for 3′-labelling require a free 3′-hydroxyl. The preferred method is the use of RNA ligase for the addition of [5′-^{32}P]pCp to the 3′ end of RNAs. An alternative method is the addition of

Table 7. Decapping and dephosphorylation.

A.   Enzymatic decapping (43)

1.   Mix 7 μl of RNA (~1–100 pmol) with 1 μl of 0.5 M NaOAc (pH 6). Incubate for 2 min at 100°C, chill on ice and spin briefly (denaturation step).
2.   Add 1 μl of 100 mM DTT, 1 μl (2.9 U) of tobacco acid pyrophosphatase and incubate for 30 min at 37°C.
3.   Add 2 μl of 0.5 M Tris–HCl (pH 8), 1 μl of alkaline phosphatase (20 mU) and incubate for 30 min at 50°C (dephosphorylation).
4.   Add 1 μl of 0.25 M potassium phosphate (pH 9.5) to terminate reaction.

Proceed to *Table 6B* for 5′-labelling.

B.   Chemical decapping (44)

The 2′- and 3′-hydroxyl groups of the cap structure (and at the 3′ end of the RNA) are oxidized with sodium periodate. The oxidized nucleoside is removed by aniline cleavage, leaving a 5′-triphosphate group.

1.   Mix 10 μl of RNA (~1–100 pmol) with 10 μl of 20 mM $NaIO_4$ and incubate for 60 min (in the dark) at room temperature (oxidation step).
2.   Add 2 μl of 2 M NaOAc (pH 5) and 50 μl of ethanol to terminate. Chill for 5 min on dry ice (or overnight at −20°C). Spin for 10 min and carefully remove and discard the supernatant. Repeat the ethanol precipitation twice and wash with 100 μl of ethanol. Dry the pellet.
3.   Dissolve in 20 μl of 165 mM aniline, 75 mM NaOAc (adjust with acetic acid to pH 4.5) and incubate for 4 h (in the dark) at room temperature (aniline cleavage).
4.   Terminate, ethanol-precipitate and wash as described above (step 2). Continue 5′-labelling following the procedures in *Tables 6* or *9* as appropriate.

Use for RNAs which cannot be labelled using the protocol in *Table 6* because of a presumed cap structure.

Table 8. Phosphate exchange reaction (45).

1.   Mix 3.5 μl of RNA (~1–100 pmol) with 2 μl of imidazole mix (250 mM imidazole–HCl (pH 6.6), 25 mM DTT, 0.5 mM spermidine, 0.5 mM Na_2EDTA, 50 mM $MgCl_2$) and 2.5 μl of 0.25 mM ADP. Add 1 μl (~100–200 μCi) of 5′-[γ-^{32}P]ATP .
2.   Add 1 μl of polynucleotide kinase (4 U) and incubate for 30 min at 37°C.
3.   Add 9 μl of 4 M ammonium acetate (pH 7) and 45 μl of ethanol to terminate. Freeze (5 min in dry ice or overnight at −20°C) and spin for 10 min. Carefully remove and discard the supernatant. Ethanol precipitation from high salt leaves most of the residual 5′-[γ-^{32}P]ATP in the supernatant.

This reaction works well only for RNAs with a 5′-monophosphate group. The 5′-phosphate is transferred to ADP and the RNA is subsequently phosphorylated with 5′-[γ-^{32}P]ATP. Direct phosphorylation of a 5′-hydroxyl group is very poor under these conditions.

Table 9. Capping with guanylyl transferase (44).

1.   Mix 3 μl of RNA (1–100 pmol) with 1 μl of 700 mM Tris–HCl (pH 8.0) and denature for 2 min at 100°C. Chill on ice and spin briefly.
2.   Add 1 μl of 100 mM DTT, 1 μl of 60 mM $MgCl_2$ and 3 μl of 5′-[α-^{32}P]GTP (30 μCi).
3.   Add 1 μl of guanylyl transferase (1 U) and incubate for 90 min at 37°C.
4.   Add 10 μl of 4 M ammonium acetate (pH 7) and 50 μl of ethanol to terminate the reaction. Freeze (5 min in dry ice or overnight at −20°C) and spin for 10 min. Carefully remove and discard the supernatant. Ethanol precipitation from high salt leaves most of the residual 5′-[α-^{32}P]GTP in the supernatant.

For this procedure, a triphosphate group is required at the 5′ terminus of the RNA, i.e. it works only with RNAs which retain the 5′ end of the primary transcript (not for capped RNAs; see *Table 7*; the chemical decapping reaction results in a 5′-triphosphate group suitable for this capping reaction).

Table 10. Addition of [5'-^{32}P]pCp with RNA ligase (46).

A. Synthesis of [5'-^{32}P]pCp (convenient and inexpensive)

1. Mix 5 μl of incubation buffer [350 mM Tris−HCl (pH 8), 30 mM MgCl$_2$, 24 mM DTT, 5 mM spermidine] with 5 μl of 1 mM Cp (cytidine-3'-phosphate) and 5 μl of 5'-[γ-^{32}P]ATP (0.5−1 mCi).

2. Add 5 μl of polynucleotide kinase (25 U; use the 3'-phosphatase-free mutant to prevent conversion of pCp to useless pC).

3. Incubate for 60 min at 37°C.

Use the radiolabelled pCp immediately or store at −20°C for up to 3 weeks.

B. Ligation reaction

1. Prepare deionized DMSO (dimethylsulphoxide) by mixing 0.5 ml of DMSO with approximately 100 mg of mixed-bed ion-exchange resin (e.g. Amberlite MB-1). Store DMSO with the resin at room temperature.

2. Mix 1 μl of RNA (~1−100 pmol) with 1 μl of DM [66% deionized DMSO, 20 mM Hepes-KOH (pH 8.3)].

3. Heat for 2 min at 100°C, chill in ice and spin briefly (denaturation step).

4. Add 2 μl of SM [120 mM Hepes-KOH (pH 8.3), 25% deionized DMSO, 10 mM DTT, 30 mM MgCl$_2$, 30 μg ml^{-1} RNase-free BSA or autoclaved gelatine, 75 μM ATP].

5. Combine with 100 μCi of [5'-^{32}P]pCp (dried down in a Speed Vac).

6. Add 2 μl of RNA ligase (6 U) and incubate for 20 h at 4°C.

7. Add 10 μl of 4 M ammonium acetate (pH 7), 4 μl of water, 50 μl of ethanol to terminate. Freeze (5 min on dry ice or overnight at −20°C) and spin for 10 min. Carefully remove and discard the supernatant. Ethanol precipitation from high salt leaves most of the residual pCp with the supernatant.

Table 11. 3' end-labelling with cordycepin triphosphate and poly A polymerase (47).

1. Mix 5 μl of RNA (1−100 pmol) with 1 μl of 100 mM Tris−HCl (pH 8) and denature (2 min at 100°C, chill on ice and spin briefly).

2. Add 1 μl of 670 mM MgSO$_4$ and 1 μl of 20 mM MnCl$_2$ and combine with 50 μCi of 5'-[α-^{32}P]3'-deoxyadenosine 5'-triphosphate (dried down).

3. Add 2 μl of poly A polymerase (4 U) and incubate for 60 min at 37°C.

4. Add 10 μl of 4 M ammonium acetate (pH 7) and 50 μl of ethanol to terminate the reaction. Freeze (5 min on dry ice or overnight at −20°C) and spin for 10 min. Carefully remove and discard the supernatant. Ethanol precipitation from high salt leaves most of the residual 5'-[α-^{32}P]3'-dATP with the supernatant.

5'-[α-^{32}P]adenosine 5'-triphosphate with poly(A) polymerase. The last method detailed here uses the highly specific addition of 5'-[α-^{32}P]adenosine 5'-triphosphate to the 3' end of tRNAs with tRNA nucleotidyl transferase.

4.5 Isolation of the labelled RNA (42)

Dissolve dried RNA from one of the preceding labelling protocols in 5−10 μl of 8 M urea/dye mix, 0.03% tracking dyes: bromophenol blue and xylene cyanol. Denature (2 min at 100°C, chill on ice, spin briefly), and load the sample on a denaturing poly-acrylamide gel (0.5 mm thick 'sequencing gel'; see Sections 4.1 and 5.5). Choose a gel concentration appropriate for your RNA: 20% for up to approximately 50 nucleotides long RNA (fragments), 8% can be used for approximately 300 nucleotides long

Table 12. 3′ end-labelling of transfer RNAs with nucleotidyl transferase (48).

A. Exonuclease treatment of tRNAs for the controlled removal of the CCA-end

1. Mix 11 μl of tRNA (~200 pmol), 2 μl of 50 mM Tris−HCl (pH 8) and 1 μl of 100 mM MgCl$_2$.
2. Add 1 μl of snake venom phosphodiesterase (0.1 μg) and incubate for 20 min at 20°C.
3. Add 30 μl of phenol and 15 μl of water and mix to terminate the reaction. Spin, remove aqueous phase and reextract it with phenol. Add 3 μl of 2M NaOAc (pH 5) and 75 μl of ethanol, chill, spin for 10 min and remove the supernatant. Wash the pellet with 100 μl of 70% EtOH and dry it *in vacuo*.

B. Restoring the CCA-end with tRNA nucleotidyl transferase

1. Add 15 μl of water to the RNA pellet from A above.
2. Mix with 5 μl of 0.25 mM CTP, 5 μl of 80 mM DTT, 2 μl of 1 M Tris−HCl (pH 8) and 20 μl of 5′-[α-^{32}P]ATP (200 μCi).
3. Add 3 μl of tRNA nucleotidyl transferase (2.5 μg μl^{-1}; 25 mU μl^{-1}) and incubate for 45 min at 37°C.
4. Add 50 μl of 4 M ammonium acetate (pH 7) and 250 μl of ethanol to terminate the reaction. Freeze (5 min in dry ice or overnight at −20°C) and spin for 10 min. Carefully remove and discard the supernatant. Ethanol precipitation from high salt leaves most residual ATP with the supernatant.

The addition is specific; tRNA nucleotidyl transferase adds only CTP and ATP to tRNA in restoration of the CCA end. Thus tRNAs are specifically labelled even in a mixture of other species of RNA. This method is also very useful for labelling tRNAs for probes in gene cloning experiments.

Table 13. RNA length relative to sequencing gel tracking dyes.

Gel (% acrylamide)	20		8		4	
Dye	BB	XC	BB	XC	BB	XC
Approximate RNA length (bases)	6	27	20	70	75	200

RNAs, whereas 4% are suitable up to 16S rRNA size (~1600 bases). Refer to *Table 13* for the relative mobilities of the tracking dyes, which serve as guidelines. *Table 14* details the recovery of RNA from polyacrylamide gels.

4.6 Troubleshooting: inefficient or no labelling of RNA

4.6.1 *Presence of inhibitor*

Inhibitors encountered include SDS, phenol and ammonium ions. Mix an otherwise well labelled RNA with your sample and check its labelling efficiency. If it is poorly labelled, then reprecipitate the sample repeatedly from ethanol, removing all the supernatant before each resuspension. If this fails to improve labelling efficiency, apply RNA to a small DEAE column. Wash with ethanol, 70% ethanol and water. Elute with 1 M NaCl or 1 M Tris−HCl. Recover by ethanol precipitation. Other column supports also should be considered (e.g. Schleicher & Schuell Elutip-r).

4.6.2 *Radiochemicals*

To check 5′-[γ-^{32}P]ATP, mix a few thousand c.p.m. with cold ATP and apply to a PEI-cellulose plate. Develop for 20 min in 0.5 M LiCl/0.5 M acetic acid. At least 90% of the radioactivity (autoradiograph) should co-migrate with ATP (position shown by fluorescence under UV light). Phosphate moves faster than ATP. To check [5′-^{32}P]-

Table 14. Recovery of RNA from polyacrylamide gels.

A. Film exposure

1. After electrophoresis, remove one glass plate and cover the gel with Saran Wrap. Fix four permanent radioactive markers at the corners and expose to X-ray film at room temperature (1 – 30 min should be appropriate to detect bands with ~5000 c.p.m.).
2. Align the film and gel on a light box. Excise radioactive bands with a scalpel and transfer to sterile Eppendorf tubes.

B. RNA elution

Select the appropriate method according to the length of the RNA.

1. Elution by diffusion

   Use soaking for RNAs up to approximately 300 bases. Add 200 μl of elution buffer (0.5 M Tris – HCl pH 7, 0.1% (w/v) SDS, 0.1 mM Na_2EDTA, 1 mM $MgCl_2$) to the gel fragment; if desired, 5 μg of carrier RNA can be added. Freeze on dry ice and agitate vigorously overnight. Remove the eluate and rinse the gel with 100 μl of elution buffer. Add 750 μl of ethanol, chill on dry ice and spin for 15 min. Remove the supernatant completely with a pipette tip and dry the pellet.

   Note: avoid the use of ammonium salts in the elution buffer, since some enzymes (e.g. ligase and kinase) are inhibited by ammonium ions. Crushing the gel prior to buffer addition does not improve elution and requires the addition of a filtration step.

2. Electroelution.

   Electroelution is required for larger RNAs. Place the gel piece into the sterile (autoclaved) Bio-trap chamber. Electroelute at 200 V for 4 h. Recover the RNA by ethanol precipitation (0.1 vol 2 M NaOAc pH 5 and 2.5 vols ethanol). Dissolve the RNA pellet in an appropriate volume of sterile water and store frozen at −20°C.

These protocols are also suitable for unlabelled nucleic acids [detected by staining (e.g. for 30 min with 1 μg ml^{-1} ethidium bromide in electrophoresis buffer visualized with long wavelength UV light) or UV shadowing (place the gel on a fluorescent thin-layer plate protected with Saran Wrap, e.g. aluminum oxide F 254 from Merck visualized with UV light)].

pCp, mix a few thousand c.p.m. with cold pCp (and ATP). Apply to a PEI-cellulose plate and develop in 0.8 M ammonium sulphate. ATP migrates slower (R_f ~0.3) than pCp (R_f ~0.6).

4.6.3 *Secondary structure effects*

Alter the reaction conditions as follows to improve labelling efficiency, e.g. inefficient labelling of a recessed terminus within duplex structure. For 5′-labelling, dephosphorylate (*Table 6A*) at 65°C and include 10% deionized DMSO in the kinase reaction (*Table 6B*). For some RNAs, the phosphate exchange reaction (*Table 8*) gives much better results. Alternatively, the recessed terminus can be exposed by controlled exonuclease digestion. A 3′-exonuclease treatment is used to expose a recessed 5′ end (*Table 12A*; using increasing amounts of enzyme). A limited digestion with a 5′-exonuclease (spleen phosphodiesterase) can be used to expose the 3′ end of RNA. Since the phosphodiesterase works only with RNA which has a free 5′-hydroxyl group, remove the terminal phosphates with phosphatase (*Table 6A*) prior to the exonuclease treatment.

Table 15. Random RNA cleavage with sulphuric acid.

1.	Add 4 μl of the incubation buffer (7 M urea, 0.11 M H_2SO_4, 0.03% dyes; *Table 16*) to the dried RNA containing 5 μg of carrier RNA.
2.	Heat for 3 min in boiling water and freeze on dry ice.

4.6.4 *Termini blocked*

Blockage may be caused by protein or if the RNA is circular. As a last resort, fragment the RNA and label the isolated fragments (19).

4.7 **Homogeneity of labelled RNAs (19)**

Examine the homogeneity of the labelled RNA before proceeding to sequence analyses. Use random acid hydrolysis (*Table 15*) to digest an aliquot of the labelled RNA (5000 – 30 000 c.p.m.). Analyze the products on a 20% sequencing gel (see Section 4.12). An example is shown in *Figure 10*.

4.8 **Overview of base-specific enzymatic digestions (42)**

Refer to *Table 16* for the partial digestion of end-labelled RNA with base-specific enzymes.

4.9 **Overview of base-specific chemical cleavage reactions (49)**

The original chemical approach was limited to 3'-labelled RNAs, since only the purine-specific reactions worked for 5'-labelled RNA (49). More recent protocols include modification with hydroxylamine at pH 10 (specific for uridine) and at pH 5 (specific for cytidine) (36,50). These reactions can be used on 5'-labelled RNAs, but the C-specific reaction results in the same mobility shift observed for 3'-labelled RNA (see Section 4.14). Zhang *et al.* (50) developed a 'solid phase' approach which avoids repeated ethanol precipitations.

Precautions. The chemicals are hazardous and the following precautions should be observed. Work in a fume hood and wear disposable gloves. Discard all solutions which contain DMS in a beaker with 10 M NaOH. Dispose of hydrazine and hydroxylamine in 3 M $FeCl_3$.

4.10 **Solutions for chemical cleavage**

(i) NaOAc. 50 mM sodium acetate (pH 4).

(ii) Prepare DMS (dimethyl sulphate), $NaBH_4$ and diethylpyrocarbonate solutions immediately before use.

(iii) 50% hydrazine. Prepare by mixing anhydrous hydrazine with an equal volume of water (stable at 4°C). This formulation is comparable to 85% hydrazine hydrate (Sigma).

(iv) 3 M NaCl in hydrazine. Prepare immediately before use with ice-cold hydrazine and NaCl dried overnight (180°C).

(v) 2 M NH_2OH (pH 10). Dissolve hydroxylamine hydrochloride in concentrated ammonia (half the calculated volume), adjust to pH 10 with ammonia and add water to the final volume. This solution is stable at −20°C.

Figure 10. Autoradiograph of randomly cleaved 5′-³²P-labelled RNAs. **Lane a** is of homogeneous RNA. Arrows indicate the large spacings characteristic for guanosines (19); **lanes b−d** show double banding characteristic of heterogeneous samples. Recommended exposure times at −70°C, using pre-flashed films and intensifying screens: for reactions containing 5000, 10 000 and greater than 20 000 c.p.m., use 50, 20 and 10 h exposure times respectively.

Table 16. Conditions for base-specific partial enzymatic digestion.

Specificity	Nuclease (amount for 5 µg of RNA)	Incubation buffer
—	—E (control)	8 M urea, 2 mM Na$_2$EDTA,
Gp/N	RNase T1 (30 mU*)	20 mM sodium citrate
ApP/N (Gp/N)	RNase U2 (60 mU*)	(pH 3.5), 0.03% dyes
Np/N	Sulphuric acid	7 M urea, 0.11 M H$_2$SO$_4$
(random)	(see *Table 15*)	0.03% dyes
Np/U + Np/A	Nuclease S7 (15 U)	8 M urea, 20 mM Tris—HCl
Cp/N (Ap/N)	RNase C13 (80 mU)	(pH 7.5), 0.03% dyes
Up/N + Ap/N	RNase Phy M (2 U)	8 M urea, 1 mM Na$_2$EDTA,
		20 mM sodium citrate
		pH 5, 0.03% dyes

A solidus indicates the cleavage specificity of each nuclease. See text for the procedure. The indicated enzyme amount serves only as a guideline.
*Note that units refer to the Sankyo product. Use 100-fold higher amounts for Boehringer-Mannheim Biochemicals unit designation.

(vi) 2 M NH$_2$OH (pH 5.5). Dissolve hydroxylamine hydrochloride in water (half the calculated volume), adjust to pH 5.5 with diethylamine (this solution is stable at −20°C). The diethylamine cannot be substituted.

(vii) 1 M aniline (pH 4.5). Mix aniline (use a fresh, only slightly yellowish solution) with water and adjust to pH 4.5 with acetic acid. The initially turbid solution will clear with acidification. If a precipitate forms, remove it by centrifugation. Store in the dark at −20°C.

4.11 Chemical cleavage reactions (36,50)

Add carrier tRNA (20 µg per reaction) and distribute to individual Eppendorf tubes (1000−50 000 c.p.m. each). Dry (desiccator or Speed Vac) and follow the procedures listed in *Figure 11*.

4.12 Polyacrylamide gel electrophoresis

The RNA samples obtained in the enzymatic protocols (Section 4.8) are already in a urea/dye solution. Resuspend dried RNA (chemical cleavage) in 5−10 µl of 8 M urea, 0.03% tracking dyes bromophenol blue and xylene cyanol. Denature the samples (2 min at 100°C, chill in ice, spin briefly) immediately before loading onto the sequencing gel (0.2−0.5 mm). Use 20% acrylamide 40−50 cm gels (for resolving up to ∼50 bases) and 6% or 8% acrylamide (for ∼120 nucleotides). Use the tracking dye migration as a guideline for running time (*Table 13*). Additional comment on gel preparation and running is given in Sections 5.2−5.5 of Chapter 2. Expose film directly to the gel after the run or fix and dry the gel prior to exposure. An intensifier screen can be used to reduce exposure time but results in more diffuse bands.

Reactions	Specificity					
	G	A (G)	U	C	U*	C*
Modification	10 μl 0.3% DMS in NaOAc	150 μl NaOAc 1 μl diethyl-pyrocarbonate	10 μl 50% hydrazine	10 μl 3 M NaCl in hydrazine	10 μl 2 M NH$_2$OH (pH 10)	10 μl 2 M NH$_2$OH (pH 5.5)
STOP and EtOH precipitation	40 sec/90°C 150 μl NaOAc 650 μl EtOH	10 min/90°C 400 μl EtOH	15 min/ice 150 μl NaOAc 450 μl EtOH	10 min/ice 500 μl 70% EtOH	40 sec/90°C 150 μl NaOAc 550 μl EtOH	7 min/90°C 150 μl NaOAc 550 μl EtOH
Resuspension and EtOH precipitation	—			150 μl NaOAc 450 μlEtOH		
EtOH wash			800 μl EtOH			
Reduction	10 μl 0.5 M NaBH$_4$ in H$_2$O 10 min/ice	—	—	—	—	—
Precipitation	150 μl NaOAc 650 μl EtOH	—	—	—	—	—
Wash	800 μl EtOH	—	—	—	—	—
Aniline cleavage	Dissolve pellet in (ca. 50 μl) 1 M aniline (pH 4.5). Incubate for 20 min at 60°C, add 150 μl of NaOAc, 650 μl of EtOH and spin for 10 min. Wash the pellet with 800 μl of EtOH and spin again. The dried pellet is dissolved in 5 μl of 8 M urea, 0.03% dyes (BB and XC), denatured by boiling and applied on a sequencing gel (Section 4.12).					

Figure 11. Reaction conditions for chemical cleavages. Add the specified modification solution to the dry RNA and incubate as indicated. Vary the reaction times to vary the amount of cleavage. The reactions are terminated by mixing with NaOAc (50 mM sodium acetate, pH 4) and ethanol. Spin for 10 min and continue with the steps as described. For the wash steps, vortex the pellets with ethanol, spin again and dry the RNA pellets. U* and C* refer to the hydroxylamine-based cleavage of these nucleotides.

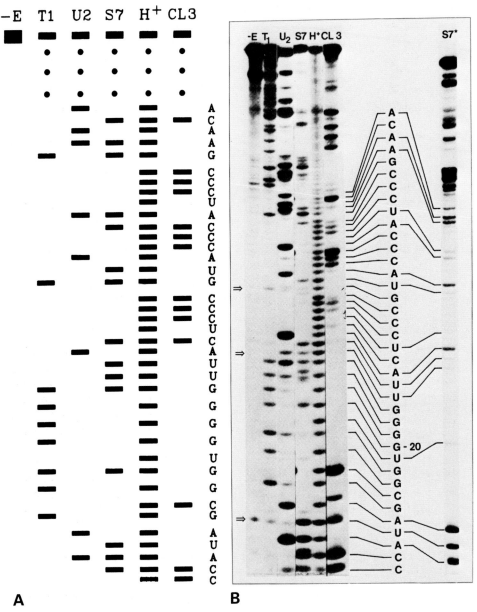

Figure 12. (**A**) Schematic representation of an idealized enzymatic RNA sequencing gel. (**B**) Autoradiograph of a sequencing gel. A 5′-^{32}P-labelled RNA was digested with the indicated enzymes: −E, no enzyme as control; T1, 30mU RNase T1; U2, 60 mU RNase U2; S7 (with 10 mM CaCl$_2$), 3 U nuclease S7; H+, 0.11 M sulphuric acid; C13, 80 mU RNase C13. The extra lane S7* shows the completely changed pattern if calcium is omitted from the reaction (15 U nuclease S7).

4.13 Interpretation of sequencing gels from enzymatically digested RNA

Each nucleotide is represented as a band in the random hydrolysis lane (H$^+$). The purines are identified by bands in the respective lanes, guanosines (lane T1) and adenosines (lane U2). Read the purines first and record the missing nucleotides as dots. For example and with reference to *Figure 12*, first record '..A.AG.GG.GGGG..A.....G .A...A....GAA.A', then identify the cytidines (lane Cl 3) and uridines (lane S7). Note that nuclease S7 cleaves *before* (5' side) U and A, whereas all other RNases cleave *after* the respective nucleotide. An idealized schematic and inferred sequence is shown in *Figure 12A*. The arrows in *Figure 12B* point to the following problem areas.

(i)   Labile bonds between a pyrimidine and adenosine: at position 13 (U) there is a band in the control ('−E') and a strong band in the 'Cl 3' digestion lane. However, the *relevant* band (S7 digestion lane) is one step below (strong band for U) and must not be confused with the band in '−E'.

(ii)   The weak band in the U2 digestion lane for U_{24} must be disregarded. Identification of this position is not possible from this gel.

(iii) Only a weak band for G_{32} in 'T1' is visible, due to local secondary structure.

(iv)   Several secondary structure effects are apparent in the S7 digestion lane. This digestion included 10 mM $CaCl_2$ as a cofactor specified in the original protocol. However, its inclusion alters the cleavage pattern through more structured regions of the RNA, apparently by stabilizing structure. Although nuclease S7 is dependent on calcium, the endogenous amount is adequate and omission of calcium results in a unambiguous digestion pattern (S7*).

4.14 Interpretation of sequencing gels from chemically cleaved RNA

In contrast to the 'bulky' enzymes, chemical modification is much less influenced by secondary structure and interpretation is relatively straightforward (*Figure 13*). However, in addition to the four base-specific reactions, a control aniline cleavage of unmodified RNA ('K') and a random acid hydrolysis ladder ('R') should be included (*Table 15*). From this ladder it can be deduced that the gap in the sequence shown in *Figure 13A* corresponds to two nucleotides ('C*'). These were later identified as dimethyl adenosines.

Hydroxylamine cleavages are shown in *Figure 13B*. These reactions also can be used for 5'-labelled RNAs. The U-specific reaction gives a clear pattern. However, interpretation of C-specific cleavages must take into account the mobility shift. To facilitate interpretation, analyse the other three bases first and use the displaced C-pattern to fill in the gaps. The C displacement also occurs with 5'-labelled RNAs.

4.15 Troubleshooting of chemical and enzymatic sequencing reactions

4.15.1 *Non-specific degradation*

For example, bands in the control lanes '−E' (no enzyme added) or 'K' (aniline treatment of unmodified RNA). The 'background bands' usually can be reduced by stronger cleavage conditions (e.g. increased enzyme or longer incubation times). Note that some modified nucleosides are sensitive to the aniline treatment; see Section 6.3.

4.15.2 *Gaps in the sequence pattern*

This can result from secondary structure which blocks enzymatic hydrolysis. Incubate at 65°C (instead of 50°C) or use the chemical approach. Alternatively, the gaps may

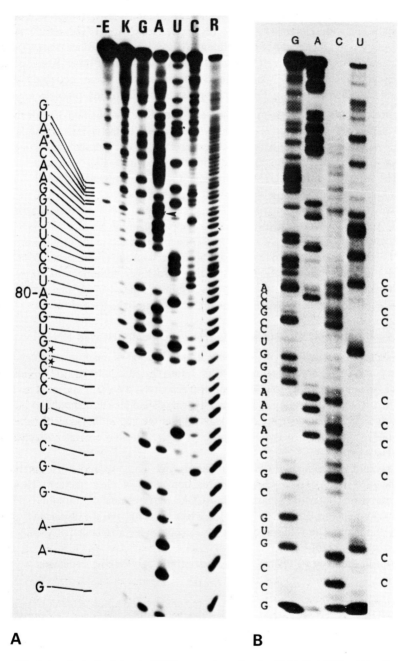

Figure 13. Autoradiograph of chemical RNA sequencing gels using 3′-labelled RNAs. (**A**) −E, untreated RNA sample; K, aniline treatment of unmodified RNA; G, dimethylsulphate; A, diethylpyrocarbonate; U, 50% hydrazine; C, anhydrous hydrazine/NaCl; R, sequence ladder (acid cleavage, *Table 15*). (**B**) 'Solid phase' analysis [courtesy of Zhang *et al.* (50)]. Similar results were obtained with reactions in solution (*Figure 11*) and also for 5′-labelled RNA. G and A reactions were as in panel A; U, hydroxylamine at pH 10; C, hydroxylamine at pH 5.5. The nucleotide sequence for lanes A, C, G, U is at the left margin, the shifted ladder for C is at the right.

correspond to modified nucleosides (see Section 6.3). The acid cleavage is insensitive to secondary structure and is blocked *only* by ribose methylations (2'-O-methyl groups).

4.15.3 *Gaps or irregular distances between bands*

First consider normal variations: the relative mobility shift with each nucleotide addition varies according to nucleotide (C < A, U < G). Anomalous mobility also may be caused by nucleoside modifications (see Section 6.3). Stable secondary structures promote 'band compression'. Incompletely denatured RNA migrates faster than less structured RNA of equal length. This results in irregular band spacings in the sequence ladder. Bands may be superimposed, rearranged or even appear as doublets. There are several approaches to resolve sequence within structured regions.

(i) *Sequence the RNA from both termini.* The proximal segment can be analysed before base pairing with the distal, complementary segment degrades gel resolution.

(ii) *Use higher temperatures during the gel run to promote denaturation.* Use diluted electrophoresis buffer (half the normal ionic strength in the gel and running buffer) to compensate for the increased conductivity resulting from the higher temperatures. In general, thinner gels (0.2 mm) give better results.

(iii) *Modify the cytidines.* Modification prevents the formation of G:C pairs (51).

5. PRIMER-EXTENSION SEQUENCING

5.1 Enzymes, buffers and reagents

(i) Avian myeloblastosis virus reverse transcriptase, Seikagaku America Inc., Boehringer Mannheim and Promega Biotec. These have been reliable sources of high-quality enzyme. Although other suppliers have been cited for sequence-grade reverse transcriptase [Molecular Genetic (SRT grade), Life Sciences (XL grade) or Pharmacia (FPLC grade)] we have had no personal experience with these.

(ii) Terminal deoxynucleotidyltransferase, Boehringer Mannheim Biochemicals.

(iii) RNase free DNase I, Promega Biotech.

(iv) Caesium trifluoroacetic acid, Pharmacia.

(v) Deoxynucleoside triphosphates 10 mM (dATP, dCTP, dGTP, dTTP), inosine and thiolated analogues [dITP and 2'-deoxyadenosine 5'-O-(α-thiotriphosphate) (α-SdATP)] Pharmacia P-L Biochemicals Inc. or Boehringer Mannheim Biochemicals. Solid nucleotides are made up in sterile 10 mM Tris−HCl, pH 7.5, the concentration determined spectrophotometrically and stored at −70°C.

(vi) 2',3'-Dideoxynucleoside 5'-triphosphates 1 mM (ddATP, ddCTP, ddGTP and ddTTP), Pharmacia P-L Biochemicals. Diluted from 10 mM stocks made up and stored as described above (Section 5.1).

(vii) 5'-[α-^{35}S]dATP, Amersham. The Dupont (NEN) product has proven equally suitable in our experience.

(viii) Standard deoxynucleotide stock. 25 μl each of 10 mM dCTP, dGTP, dTTP and 6.25 μl of 10 mM α-SdATP. Make to 100 μl final volume with 18.75 μl of 10 mM Tris−HCl, pH 7.5. (For 5'-[α-^{35}S]dATP sequencing reactions).

(ix) Inosine deoxynucleotide stock. Prepare as stock standard deoxynucleotide solution with replacement of dGTP with dITP.

(x)   Standard deoxynucleotide and inosine deoxynucleotide solutions. Made with 11.2 μl of stock solution and 88.8 μl of 10 mM Tris$-$HCl pH 8.0.

(xi)   5 × RT buffer, 0.25 M Tris$-$HCl, pH 8.5 (21°C), 0.25 M KCl, 0.05 M dithiothreitol, 0.05 M MgCl$_2$.

(xii)   5 × hybridization buffer, 0.5 M KCl, 0.25 M Tris$-$HCl, pH 8.5.

(xiii) Chase mix. 1 mM dNTPs, 6 mM Tris$-$HCl, pH 8.5

(xiv)   Stop mix. 5 mM EDTA, pH 7.2, 0.05% bromophenol blue, 0.05% xylene cyanol, 92% formamide (v/v).

(xv)   Reverse transcriptase dilution buffer. 50 mM Tris$-$HCl, pH 8.4, 2 mM DTT, 50% glycerol (v/v).

(xvi)   10 × NNB: 1.34 M Tris base, 0.45 M H$_3$BO$_3$, 25 mM Na$_2$EDTA.

(xvii) TE buffer: 10 mM Tris$-$HCl, pH 7.4, 1.0 mM Na$_2$EDTA, pH 7.4.

(xviii) Gel fix: 10% methanol, 1.5% acetic acid, 2% glycerol (v:v:w) in water.

5.2 **RNA template**

The RNA template must be intact and of good purity. Standard techniques of phenol extraction and ethanol precipitation generally yield template of adequate quality (52). The importance of a high quality template cannot be overemphasized. Store RNA at $0.5-5$ mg ml^{-1} in water at -70°C. If priming is poor and/or sequencing-gel background high and altered reaction conditions have been explored, (*Tables 18* and *19*), evaluate alternative isolation or secondary purification steps. Secondary purifications that have been demonstrated to improve ribosomal RNA templated sequencing reactions include salting out and pelleting through caesium trifluoroacetic acid (CsTFA) gradients (*Table 17*). CsTFA is a potent chaotrope. Other interfering materials (e.g. poly-saccharides) also seem to be largely removed by this procedure (*Table 17*).

5.3 **Primer design and storage**

The primers are synthetic DNA oligonucleotides containing free 3'-hydroxyl groups. Although a minimum primer length based on probability considerations is from 11 to 12 nucleotides, in practice most range from 15 to 20 nucleotides. However, even within this size range priming at multiple or alternative sites sometimes occurs. If possible, primers should be of greater than 50% G/C content so they will compete well with secondary structure within the template and they should have one or more C or G residues near the 3' (priming) end so that they 'lay down' well on the template (18). Extensive primer-extension sequencing studies of ribosomal RNAs have shown considerable variation in priming efficiency between primers of similar size and composition, presumably reflecting competing template structure. In these instances, 'sliding' the target site several nucleotides to the 5' or 3' direction has sometimes improved priming efficiency. Thus, primer design remains largely empirical.

Store primers at -70°C in sterile distilled water ($250-500$ μg ml^{-1}). Dilute at $5-25$ μg ml^{-1} for working solutions and store at -20°C.

5.4 **Hybridization and extension reactions**

The standard conditions for hybridization and primer extension are given in *Tables 18* and *20*. For convenience in adjusting the ratio of dideoxynucleotide to deoxynucleotide

Table 17. Secondary purification of RNA template.

A.   Precipitation from high salt (52)

1.   Adjust RNA concentration to approximately 5 mg ml^{-1} in 10 mM TE buffer (by dilution or resuspension following ethanol precipitation).
2.   Add an equal volume of RNA and 4.0 M NaCl to an Eppendorf tube, mix and hold overnight on ice.
3.   Pellet the high molecular weight RNA by centrifugation for 5−10 min in a microcentrifuge.
4.   Resuspend the RNA pellet in water or TE and determine the concentration spectrophotometrically (assume that a 50 μg ml^{-1} RNA solution has an A_{260} of 1).
5.   Remove residual NaCl from the RNA preparation by adding 1/10th vol of 2.0 M NaOAc and then 2.0 vols of ethanol. Mix. Chill to −20°C for 4 h or more. Pellet and resuspend as in steps 3 and 4 (above).

B.   Pelleting through caesium trifluoroacetic acid

This protocol is devised for use with the Beckman TL-100 bench top ultracentrifuge for the recovery of ribosomal RNA (W.Weisburg, personal communication). Volumes and spin times must be adjusted for use with other rotors and centrifuges. Since rRNA and mRNA may differ slightly in density, average density (or run time) of the gradient may need to be adjusted accordingly.

1.   Resuspend RNA pellet in 1.4 ml of 10 mM Tris−HCl, pH 7.4, 5 mM MgCl$_2$ and 30 mM KCl.
2.   Add 10 μl of RNase-free DNase I (1 unit μl^{-1}, Promega Biotech) and incubate for 5 min at room temperature (this step is optional, depending upon the amount of contaminating DNA).
3.   Make to 1.65 g ml^{-1} by addition of 2.19 ml of CsTFA (2.04 g ml^{-1}, Pharmacia).
4.   Seal in 13 × 32 mm Beckman polyallomer bell-top quick seal centrifuge tubes.
5.   Centrifuge at 60 000 r.p.m. for 5 h or longer in a TLA 100.3 rotor at 10°C.
3.   Resuspend the pellet in sterile distilled water and reprecipitate from ethanol.

These procotols have yielded improved ribosomal RNA template but have *not* been examined for other RNAs.

Table 18. Hybridization of primer and template.

---

1.   Combine in a 0.5-ml Eppendorf tube: 2 μl of primer (5−25 μg ml^{-1}), 2 μl of RNA template (~0.5 mg ml^{-1}), and 1 μl of 5 × hybridization buffer.
2.   Heat in a dry block heater with water surrounding the tube at 90°C for 1 min. [Heating to 65°C is usually equally effective (52).]
3.   Remove dry block from heating unit and allow to cool to about 40°C over 10−20 min.
4.   Centrifuge the tube briefly to bring the condensate to the bottom of the tube and hold it on ice until adding to the extension reaction.

Initially use the lower amount of primer indicated. The ratio of primer to template should be about 1:1. If priming is inefficient, as indicated by weak bands and long exposure times, increase the amount of primer several fold. Shadow banding suggests priming at multiple sites. Priming at secondary sites sometimes can be eliminated, or reduced, by increasing the stringency of the hybridization. Increase the stringency by substituting water for the 5 × hybridization buffer.

---

in the analogue mixes, prepare two stock solutions. The first (standard deoxy or inosine deoxy) contains only the deoxynucleotides at the same concentration as the dideoxy solution (*Table 19*). Adjust the ratio of dideoxy- to deoxynucleotides by mixing the two solutions. The ratio should be adjusted such that most labelled chains are terminated within the resolving power of the gel (*Figure 14*). The deoxy-solution containing the helix-destabilizing inosine substituting for guanosine is used to resolve sequencing gel band compression arising from undenatured secondary structure (*Figure 15*).

Table 19. Formulation of dideoxynucleotide and analogue solutions.

A. Dideoxynucleotide solutions (μl)

	ddA	ddA_i	ddC	ddC_i	ddG	ddG_i	ddT	ddT_i
Standard deoxynucleotide stock (see Section 5.1, viii)	11.2	–	11.2	–	11.2	–	11.2	–
Inosine deoxynucleotide stock (see Section 5.1, ix)	–	11.2	–	11.2	–	11.2	–	11.2
0.1 mM ddA	12.0	6.0	–	–	–	–	–	–
1.0 mM ddC	–	–	12.0	6.0	–	–	–	–
1.0 mM ddG	–	–	–	–	17.0	3.0	–	–
1.0 mM ddT	–	–	–	–	–	–	35.0	12.0

Make to 100 μl final volume with 10 mM Tris, pH 8.0.

B. Analogue solutions

> 1 part dideoxynucleotide solution
> 6 parts standard deoxynucleotide or inosine solutions.

Use this as a starting formulation. Adjust the average chain length by altering the proportion of dideoxynucleotide solution to deoxynucleotide solution.

Final μM concentrations for analogue solutions at 1 part to 6 parts formulation:

> Standard ddN/dN: C, 17/250; A, 1.7/62.5; T, 50/250; G, 24/250
> Inosine ddN/dN: C, 8.6/250; A, 0.86/62.5; T, 17/250; I, 4.3/250.

Table 20. Extension reaction.

Use a P20 Pipetman or equivalent for dispensing all of the following.

1. Label five 0.5-ml Eppendorf tubes (C_1, A, T, G, C_2) and aliquot 5 μCi [α-^{35}S]dATP to each. For convenience and reproducibility, dilute dATP to a volume convenient for dispensing (e.g. 5 μl per tube). Dry in a SpeedVac (Savant Instruments).
2. While ATP is drying, combine in a 0.5-ml Eppendorf tube: 5.5 μl of 5 × RT buffer, 5.5 μl of primed template, and 5.5 μl of reverse transcriptase (~ 1000 units ml^{-1}).
3. Mix gently and dispense 3 μl to the bottom of each [^{35}S]dATP-containing Eppendorf tube.
4. Add 2 μl of the designated analogue mix to the upper side wall of each tube and combine with enzyme mix at the bottom of the tube by tapping the rack containing the tubes and by gentle mixing or alternatively by brief centrifugation. A control reaction which contains only the four dNTPs (no ddNTP) may also be included as a sixth reaction in this series.
5. Incubate for 5 min at room temperature and 30 min at either 37°C or at 55°C[a].
6. Add 1 μl of chase mix, mix gently and continue incubation at 37°C for 15 min.
7. Add 10 μl of stop mix, vortex and store at −20°C until loading the sequencing gel. Run the sequencing gel within a week for best results. Reactions are heated to either 90°C or 65°C immediately prior to loading. The less harsh 65°C incubation is usually as effective as the 90°C incubation (52).

[a]Higher incubation temperature (55°C) sometimes alleviates premature termination by the reverse transcriptase at certain points in the template. These appear as a band at the same position in all tracks obscuring the genuine stop. Deoxynucleotidyltransferase (TdT) is also used to eliminate obscuring prematurely truncated transcripts (*Table 21*).

5.5 Polyacrylamide gel preparation, electrophoresis and film exposure

Sequencing gels are made as detailed elsewhere in this volume (Section 4.12 and Chapter 2, Section 5.2). The description here is limited to overview and referencing

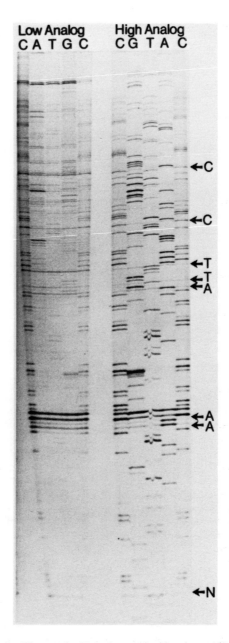

Figure 14. Effect of changing dideoxynucleotide to deoxynucleotide ratios on 16S rRNA-templated sequencing reactions. The μM ratios for the high and low analogue formulations are as follows: high ddN/dN, C = 60/250, A = 3.5/65, T = 80/250, G = 50/250, low ddN/dN, C = 10/250, A = 0.5/65, T = 15/250, G = 10/250. The low analogue reactions contained a suboptimal ratio. Increasing the dideoxynucleotide concentration resolved many obfuscating termination sites. The position indicated by N is 11 nucleotides from the 3′ terminus of the primer. Note that the order of loading is different for the two gels.

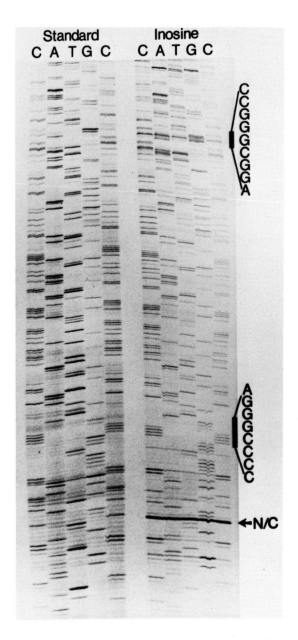

Figure 15. Sequencing gels of guanosine (standard) versus inosine substituted 16S rRNA-templated sequencing reactions. The two designated regions of compression within G/C-rich regions were resolved by inosine substitution. The N/C marks a position of premature termination resulting from inosine inclusion.

unless there is considerable variation from standard protocols. Prepare $0.5 \times 2.5 \times$ NNB buffer gradient gels as described by Biggin *et al.* (53) and run gels with $0.5 \times$ reservoir buffer (52). For gels 0.3 mm thick, 18 cm wide and 49 cm long, run at $35-50$ W on an apparatus having no heat sink (constant power).

(i) *Polymerization*. Immediately after pouring, lay the gel flat and insert comb. Place a strip of Whatmann 3 MM paper over the comb and saturate with 0.5 × buffer acrylamide solution and allow the gel to polymerize undisturbed for 30 min. The filter strip prevents loss of acrylamide from the top of the gel (via capillary action and siphoning) that causes uneven polymerization near the loading surface and loss of resolution. Gels can be run within an hour or two after polymerization.

(iii) *Fixation and drying*. Remove the 'top' plate and secure the gel to the lower plate with rubber bands around the top and lower region of the gel and plate. These secure the gel to the plate (without cutting it) for ease of handling during fixation. Soak the gel in gel fix for one or more hours (or overnight), remove the rubber bands and transfer the gel to Whatman 3MM paper. Dry on a slab gel drier and expose to X-ray film (3−4 days or as necessary).

5.6 **Resolution of sequencing gel ambiguities and troubleshooting**

Sequence information is commonly lost as a result of two different anomalies, band compression and premature termination of the transcript by reverse transcriptase. Band compression is a consequence of undenatured secondary structure (usually G/C-rich hairpin helices). The inclusion of helix-destabilizing analogues (inosine or deazaguanine) frequently eliminates or reduces this effect (*Figure 15*). Formulation of an inosine-containing reaction mix is given in *Table 19*. Because inosine tends to promote premature termination of transcripts it is only generally used for secondary 'clean-up' sequencing reactions.

Premature termination of the transcript results in a band across all lanes of the sequencing gel, obscuring the identity of the dideoxynucleotide truncation. Coincident termination is the consequence of nicked template (generally caused by RNase), secondary structure or modified nucleotides. Since the obscuring transcripts still retain 3′-hydroxyl groups, they are substrates for chain elongation and can be chased out (of a unique gel position) by chain extension with terminal deoxynucleotidyl transferase (TdT) (54). The TdT amended sequencing reaction is given in *Table 21*. Since some spurious banding can result from the inclusion of the TdT extension step, this too should be used as a secondary clean-up reaction.

6. IDENTIFICATION OF MODIFIED NUCLEOSIDES BY 2-D CHROMATOGRAPHY

6.1 **Equipment and suppliers**

Only special items are listed here, as most materials are listed in Section 4.

(i)   Cellulose thin layer plates, Merck.
(ii) Chromatography chambers, Desaga.
(iii) Nucleotides, Pharmacia, Boehringer-Mannheim Biochemicals or Sigma Chemical Company.
(iv) Apyrase, grade III, Sigma Chemical Company.

6.2 **Analysis of the nucleoside composition (6)**

The RNA is digested completely with RNase T2, labelled (if necessary) and the

Table 21. Terminal deoxynucleotidyltransferase (TdT) chase of sequencing reactions.

Follow standard or inosine substituted sequencing reaction through 30 min incubation, then:

1. Boil reaction tubes.
2. Spin briefly to bring condensate down and add 1 μl of TdT and 1 μl of standard chase mix (several concentrations of TdT should be examined in preliminary trials, e.g. 3.5 and 7.0 units). The enzyme is diluted with the recommended storage buffer immediately prior to use. We have observed loss of activity following storage of diluted enzyme.
3. Incubate for 30 min at 50°C (50°C incubation was slightly better than 37°C incubation in one case examined).
4. Add stop mix.

Table 22. Complete digestion with RNase T2 and phosphorylation of the nucleotides.

1. Mix 5 μl (0.1−0.5 μg) of purified RNA with 2 μl of 50 mM ammonium acetate (pH 4.5).
2. Add 3 μl of enzyme (use a mixture of 50 mU RNase T2 and 100 ng RNase A) and incubate for 4 h at 37°C.
3. Dry in a desiccator.
4. Dissolve the digested RNA in 5 μl of water and add 1 μl of 1 M Tris−HCl (pH 8.0), 1 μl of 0.2 M MgCl$_2$, 32 mM spermidine, 1 μl of 100 mM DTT and 1 μl of 5′-[γ-^{32}P]ATP (50 μCi; diluted with cold ATP to 300 Ci mmol^{-1}).
5. Add 1 μl of polynucleotide kinase (5 U) and incubate for 60 min at 37°C.
6. Remove excess ATP by adding 1 μl of apyrase (5 U; grade III from Sigma) and incubating for 20 min at 37°C.

Table 23. Nuclease P1 digestion.

1. Mix 3 μl of the digest (*Table 22*) with 1 μl of 750 mM ammonium acetate (pH 5.3) and 5 μl of water.
2. Add 1 μl of nuclease P1 (2 μg). Incubate for 3 h at 37°C and store at −20°C.

This treatment removes the 3′-phosphate groups and at least partially cleaves dinucleotides ^{32}pN(m)pN (see also *Table 25*).

nucleotides identified by two-dimensional thin layer chromatography. Although HPLC methods (29) give superior and quantitative data, this analysis requires at least 10-fold higher amounts of RNA ($\sim$5 μg). Only the thin-layer analysis will be described here.

(i) Preparation of labelled nucleotides from RNAs. RNA labelled *in vivo* is directly digested (*Table 22*) and fractionated by thin layer chromatography. Non-radioactive RNA digestion products [nucleoside 3′-phosphates or dinucleotides resistant to RNase T2; N(m)pNp] are 5′-^{32}P-phosphorylated with polynucleotide kinase and 5′-[γ-^{32}P]ATP (*Table 22*). Analyse the nucleoside diphosphates directly (using the system of Gupta and Randerath, 55) or alternatively, following removal of the 3′-phosphates with nuclease P1 (*Table 23*). Include a control (e.g. 5S rRNA without modified nucleosides or a sample without any RNA) to identify spurious spots (products or contaminants) derived from the radiochemicals.

(ii) Two-dimensional thin layer chromatography (56). Dry 4 μl of the sample (*Table 23*) in a desiccator and dissolve in 1 μl of a nucleotide mixture ($\sim$5 mg ml^{-1} each of pA, pC, pG and pU; include other standards if appropriate). Apply at the lower left corner on a cellulose thin layer plate and dry with a hair dryer.

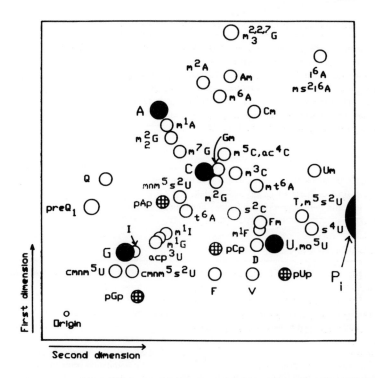

Figure 16. Schematic nucleotide positions in the thin layer system of Nishimura (56). First dimension in isobutyric acid/conc. ammonia/water: 57.7/3.8/38.5 (solvent A), second dimension in isopropanol/conc. HCl/water: 70/15/15 (solvent B). Note that relative mobilities can vary and depend on the age of the solvent.

Before separating the radioactive sample, it is a good idea to perform a test run for each solvent, using the nucleotide mix. The reference nucleotides can be located under UV light (254 nm).

(iii) Develop in solvent A (isobutyric acid/conc. ammonia/water: 57.7/3.8/38.5; pH 4.3) and dry overnight in a fume hood.

(iv) Turn the plate and develop in solvent B (isopropanol/conc. HCl/water: 70/15/15). Dry in a fume hood. Fix permanent radioactive markers and expose to X-ray film.

Use *Figure 16* as a guide for nucleotide assignment. Co-chromatography with an authentic sample is required for positive identification. Elute nucleotides of interest (*Table 24*), mix a reference nucleotide and use the same or another thin layer system (6,55) to confirm identity. Use more rigorous digestion (*Table 25*) to cleave dinucleotides (pN(m)pN).

6.3 Rapid detection of modified nucleosides (30)

The methods for sequencing end-labelled RNAs can be used for the preliminary identification of modified nucleosides. Detection and initial identification is based upon differences in sensitivity (30) of modified residues to base-specific enzymatic and chemical cleavage. A systematic survey of this behaviour was given in *Figure 8*.

Table 24. Elution of nucleotides from thin layer cellulose plates.

1.	Plug a thin sterile glass tube (i.d. ~3 mm, 3 cm long) at one conical end with sterile cotton wool.
2.	Connect to an aspirator and collect material scraped from the plate by suction.
3.	Place the tube into a shortened pipette tip fitted through the lid (punch a hole in it) of an Eppendorf tube.
4.	Add 150 μl of water to the glass tube and collect it in the Eppendorf tube by centrifugation (use a low speed swing-out rotor, e.g. with the Beckman TJ-6 table top centrifuge).
5.	Repeat with another 150 μl of water. Dry in a Speed Vac or desiccator.

Table 25. Exhaustive nuclease P1 digestion.

1.	Dissolve the sample (from *Table 24*) in 3 μl of water, add 1 μl of 100 mM ammonium acetate (pH 5.3).
2.	Add 1 μl of nuclease P1 (2 μg) and incubate for 7 h at 65°C.
3.	Dry and mix with appropriate reference nucleotides for further analysis.

Table 26. Random RNA digestion.

1.	Transfer 1 μl of RNA (0.1−0.5 μg) to a glass capillary tube (~ 10 μl capacity) and seal the ends in a flame.
2.	Immerse in boiling water for 30 sec and chill on ice.
3.	Open the capillary and transfer the hydrolysate to an Eppendorf tube containing 5 μl of water. Rinse the capillary with this solution.

Each RNA must receive a single nick. The original protocol of Stanley and Vassilenko used a formamide cleavage (31). This has the disadvantage of requiring an ethanol precipitation prior to phosphorylation to remove inhibitory formamide. Gupta and Randerath (55) introduced the cleavage in water described here. This procedure works only in glass capillary tubes (not siliconized) and not in plastic Eppendorf tubes.

6.4 A different approach for RNA sequencing: isolation and identification of modified nucleosides from sequencing gels (31)

This approach combines gel electrophoresis and thin layer chromatography to determine nucleotide sequence. Although applicable to any RNA that can be end-labelled, the approach is particularly useful for simultaneously determining sequence and the identity of modified nucleotides within the sequence. In addition, this approach produces a series of internal end-labelled RNA fragments which can be sequenced (Sections 3.2.6 and 4) to resolve ambiguities resulting from the influence of secondary structure on sequencing gel mobility.

The unlabelled RNA is randomly hydrolysed under conditions that favour a single break per strand (*Table 26*). Subsequently the 5'-hydroxyl groups of the fragments are 5'-^{32}P-labelled (*Table 27*). The 5' end of the intact RNA must be blocked (e.g. by the presence of a naturally occurring phosphate) so only those fragments having the 3' terminus are labelled. The labelled fragments are separated on a long sequencing gel. Each band of the resulting sequence ladder contains a fragment with the 5' label at that position in the sequence. Individual fragments are eluted and their 5' end determined (*Table 28*). A more rapid variation of this approach is the direct transfer of the fragments from the gel onto thin layer plates. The fragments are digested directly on the plate followed by chromatography of the resulting nucleotides (55,57). Common problems and suggested resolutions follow.

Table 27. 5'-^{32}P-Phosphorylation and isolation of the RNA fragments.

1.	Mix 6 μl of RNA hydrolysate (*Table 26*) with 1 μl of 1 M Tris−HCl (pH 8), 1 μl of 0.2 M MgCl$_2$ and 32 mM spermidine, 1 μl of DTT and 1 μl of 5'-[γ-^{32}P]ATP ($\sim$ 100 μCi; 3000 Ci mmol^{-1}).
2.	Add 1 μl of polynucleotide kinase (5 U) and incubate for 30 min at 37°C.
3.	Transfer to a dried (desiccated) 10 μl aliquot of urea/dye mix (8 M urea, 0.03% tracking dyes).
4.	Denature (100°C for 2 min and chill on ice) and load onto a long 15% or 20% sequencing gel (20 × 80 × 0.05 cm).
5.	Electrophorese at 80 W ($\sim$4000 V) until bromophenol blue has migrated approximately 60 cm. Elute the RNA fragments (*Table 14*; include 10 μg of carrier RNA) and ethanol precipitate.

Table 28. Analysis of the 5' terminal nucleosides.

1.	Use an aliquot of each fragment (*Table 27*; $\sim$ 1000 c.p.m.) in 8 μl of water. Mix with 1 μl of 500 mM ammonium acetate (pH 5.3).
2.	Add 1 μl of nuclease P1 (50 ng μl^{-1}) and incubate for 120 min at 50°C.
3.	Dry in a desiccator, mix with 2 μl of nucleotide mix (5 mg ml^{-1} each of pA, pC, pG, pU) and apply aliquots (100 c.p.m. are detected after overnight exposure) on two cellulose thin layer plates.
4.	The plates are developed separately in solvent A or B (Section 6.2) respectively.

Secondary analyses may be required for a positive identification, as discussed in Section 6.2. The occurrence of several spots from one RNA fragment is possible and may be due to incomplete digestion with nuclease P1. Repeat the analysis with another aliquot using rigorous conditions (*Table 25*).

6.5 Troubleshooting

6.5.1 *All (many) RNA fragments have the same nucleoside (adenosine) as the 5' end group*

This is due to overdigestion, where the most labile bonds (pyrimidine−adenosine) are cleaved and secondary cleavage products occur (more than one nick per molecule). This problem is rare with 'water cleavage', and shortened cleavage times should help. It is also possible that the 5' end of your intact RNA is not protected (no 5'-phosphate), especially if nucleosides other than A are observed. Check if the RNA is 5'-labelled without prior phosphatase treatment. Unfortunately, attempts to block the 5' end by complete phosphorylation with polynucleotide kinase and an excess of cold ATP have in our experience given unsatisfactory results, but improvements are possible.

6.5.2 *Only fragments from the top of the gel yield interpretable results*

Shorter fragments are uninterpretable beyond a certain point in the ladder. The RNA probably contains a labile nucleoside, where predominant cleavage occurs (e.g. mam^{5}s^{2}U). As a consequence, the bands from the gel top to the position of the labile base are weak (from 'intact' RNA). Banding intensity increases below the position of the labile nucleotide in the gel. Examine alternative fragmentation protocols, e.g. formamide (31), acid with 20 mM sulphuric acid for 30 sec in boiling water or use the protocol in *Table 15* with subsequent dilution with water (to 20 μl) and ethanol precipitation with 2 μl of 2 M sodium acetate (pH 5) and 50 μl of ethanol.

6.5.3 *Very uneven band intensities in the ladder*

The result of secondary structure which excludes the kinase. Bands may be so faint that a whole region of the gel is essentially blank. Try other fragmentations as suggested above (6.5.2).

6.5.4 *More than one nucleoside occurs per RNA fragment*

First exclude the trivial explanation of partial nuclease P1 digestion (see *Table 28*). Other possible explanations include those listed above (6.5.1 and 6.5.2) and altered electrophoretic mobility resulting from secondary structure (Section 4.15). Use the other sequencing approaches described to resolve these ambiguities. Note that the $5'-^{32}P$-labelled RNA fragments isolated from each band in the ladder are suitable for alternative methods of sequence analysis (Sections 3 and 4).

7. REFERENCES

1. Holley,R.W. (1977) In *Nobel Lectures in Molecular Biology 1933 – 1975*. Elsevier North Holland Inc., New York, p. 285.
2. De Wachter,R. and Fiers,W. (1982) In *Gel Electrophoresis of Nucleic Acids*. Rickwood,D. and Hames,B.D. (eds), IRL Press, Oxford, p. 77.
3. Sanger,F., Brownlee,G.G. and Barrell,B.G. (1965) *J. Mol. Biol.*, **13**, 373.
4. Uchida,T., Bonen,L., Schaup,H.W., Lewis,B.J., Zablen,L. and Woese,C.R. (1974) *J. Mol. Evol.*, **3**, 63.
5. Brownlee,G.G. and Sanger,F. (1969) *Eur. J. Biochem.*, **11**, 395.
6. Silberklang,M., Gillum,A.M. and RajBhandary,U.L. (1979) In *Methods in Enzymology*. Wu,R. and Grossman,L. (eds), Academic Press Inc., London and New York, Vol. 59, p 58.
7. Domdey,H. and Gross,H.J. (1979) *Anal. Biochem.*, **49**, 346.
8. Brownlee,G.G. (1972) *Determination of Sequences in RNA, Laboratory Techniques*. Work,T.S. and Work,E. (eds), North-Holland and American Elsevier, Amsterdam and New York.
9. Woese,C.R., Sogin,M., Stahl,D., Lewis,B.J. and Bonen,L. (1976) *J. Mol. Evol.*, **7**, 197.
10. Jay,E., Bambara,R., Padmanabhan,R. and Wu,R. (1974) *Nucleic Acids Res.*, **1**, 331.
11. Stackebrandt,E., Ludwig,W. and Fox,G.E. (1985) In *Methods in Microbiology*. Gottschalk,G. (ed.), Academic Press, London, Vol. 18, p. 75.
12. La Torre,J.L., Underwood,B.O., Lebendiker,M., Gorman,B.M. and Brown,F. (1982) *Infect. Immunol.*, **36**, 142.
13. Stahl,D.A., Lane,D.J., Olsen,G.J. and Pace,N.R. (1984) *Science*, **224**, 409.
14. Fox,G.E., Stackebrandt,E., Hespell,R.B., Gibson,J., Maniloff,J., Dyer,T.A., Wolfe,R.S., Balch,W.E., Tanner,R.S., Magrum,L.J., Zablen,L.B., Blakemore,R., Gupta,R., Bonen,L., Lewis,B.J., Stahl,D.A., Leuhrsen,K.R., Chen,K.N. and Woese,C.R. (1980) *Science*, **209**, 457.
15. Costa Giomi,M.P., Gomes,I., Tiraboschi,B., Auge de Mello,P., Bergman,I.E., Scodeller,E.A. and La Torre,J.L. (1988) *Virology*, **162**, 58.
16. Stahl,D.A., Lane,D.J., Olsen,G.J. and Pace,N.R. (1985) *Appl. Environ. Microbiol.*, **49**, 1379.
17. Brown,D.D. and Gurdon,J.B. (1977) *Proc. Natl. Acad. Sci. USA*, **74**, 2064.
18. Stern,S., Moazed,D. and Noller,H.F. (1989) In *Methods in Enzymology*. Wu,R. (ed.), Academic Press Inc., London and New York, in press.
19. Gross,H.J., Krupp,G., Domdey,H., Raba,M., Jank,P., Lossow,C., Alberty,H., Ramm,K. and Sänger,H.L. (1982) *Eur. J. Biochem.*, **121**, 249.
20. Shibahara,S., Mukai,S., Nishikara,T., Inoue,H., Ohtsuka,E. and Morisawa,H. (1987) *Nucleic Acids Res.*, **15**, 4403.
21. Brownlee,G.G. and Cartwright,E.M. (1977) *J. Mol. Biol.*, **114**, 93.
22. Baralle,G.R. (1977) *Cell*, **12**. 1085.
23. Proudfoot,N.J. (1977) *Cell*, **10**, 559.
24. Hamlyn,P.H., Gait,M.J. and Milstein,C. (1981) *Nucleic Acids Res.*, **9**, 4485.
25. Robbins,P.F., Rosen,E.M., Haba,S. and Nisonoff,A. (1986) *Proc. Natl. Acad. Sci. USA*, **83**, 1050.
26. Lane,D.L., Pace,B., Olsen,G.J., Stahl,D.A., Sogin,M.L. and Pace,N.R. (1985) *Proc. Natl. Acad. Sci. USA*, **82**, 6955.
27. Woese,C.R. (1987) *Microbiol. Rev.*, **51**, 221.
28. Stern,S., Wilson,R.C. and Noller,H.F. (1986) *J. Mol. Biol.*, **192**, 101.
29. Gehrke,C.W. and Kuo,K.C. (eds) (1989) *Chromatographic and Other Analytical Methods in Nucleic Acids Modification Research*. Elsevier, Amsterdam, in press.
30. Lankat-Buttgereit,B., Gross,H.J. and Krupp,G. (1987) *Nucleic Acids Res.*, **15**, 7649.

31. Stanley,J. and Vassilenko,S. (1978) *Nature*, **274**, 87.
32. Diamond,A. and Dudock,B. (1983) In *Methods in Enzymology*. Wu,R., Grossman,L. and Moldave,K. (eds), Academic Press Inc., London and New York, Vol. 100, p. 431.
33. Laskey,R.A. and Mills,A.D. (1977) *FEBS Lett.*, **82**, 314.
34. Sogin,M.L., Pace,N.R., Rosenberg,M. and Weisman,S.M. (1976) *J. Biol. Chem.*, **251**, 3480.
35. Domdey,H., Jank,P., Sänger,H.L. and Gross,H.J. (1978) *Nucleic Acids Res.*, **5**, 1221.
36. Waldmann,R., Gross,H.J. and Krupp,G. (1987) *Nucleic Acids Res.*, **15**, 7209.
37. Ehresmann,C., Baudin,F., Mougel,M., Romby,P., Ebel,J.-P. and Ehresmann,B. (1987) *Nucleic Acids Res.*, **15**, 9109.
38. Pinck,M., Fritsch,C., Ravelonandro,M., Thivent,C. and Pinck,L. (1981) *Nucleic Acids Res.*, **9**, 1087.
39. Krämer,A., Keller,W., Appel,B. and Lührmann,R. (1984) *Cell*, **38**, 299.
40. Donis-Keller,H. (1979) *Nucleic Acids Res.*, **7**, 179.
41. Branlant,C., Krol,A. and Ebel,J.-P. (1981) *Nucleic Acids Res.*, **9**, 841.
42. Krupp,G. and Gross,H.J. (1983) In *The Modified Nucleosides in Transfer RNA II: A Laboratory Manual of Genetic Analysis, Identification and Sequence Determination*. Agris,P.F. and Kopper,R.A. (eds), Liss, New York, p. 11.
43. Efstratiadis,A., Vournakis,J.N., Donis-Keller,H., Chaconas,G., Dougall,D.K. and Kafatos,F. (1977) *Nucleic Acids Res.*, **4**, 4165.
44. Pedersen,N., Hellung-Larsen,P. and Engberg,J. (1985) *Nucleic Acids Res.*, **13**, 4203.
45. Berkner,K.L. and Folk,W.R. (1977) *J. Biol. Chem.*, **252**, 3176.
46. England,T.E., Bruce,A.G. and Uhlenbeck,O.C. (1980) In *Methods in Enzymology*. Grossman,L. and Moldave,K. (eds), Academic Press Inc., London and New York, Vol. 65, p. 65.
47. Beltz,W.R. and Ashton,S.H. (1982) *Fed. Proc.*, **41**, 1450.
48. Schön,A., Krupp,G., Gough,S., Berry-Lowe,S., Kannangara,C.G. and Söll,D. (1986) *Nature*, **322**, 281.
49. Peattie,D.A. (1979) *Proc. Natl. Acad. Sci. USA*, **76**, 1760.
50. Zhang,Y., Liu,W., Feng,Y. and Wang,T.P. (1987) *Anal. Biochem.*, **163**, 513.
51. Mazo,A.M., Mashkova,T.D., Avdonina,T.A., Ambartsumyan,N.S. and Kisselev,L.L. (1979) *Nucleic Acids Res.*, **7**, 2469.
52. Lane,D.J., Field,K.J., Olsen,G.J. and Pace,N.R. (1989) In *Methods in Enzymology*. Academic Press Inc., London and New York, in press.
53. Biggin,M.D., Gibson,T.J. and Hong,G.F. (1983) *Proc. Natl. Acad. Sci. USA*, **80**, 3963.
54. DeBorde,D.C, Naeve,C.W., Herlocher,M.L. and Maassab,H.F. (1986) *Anal. Biochem.*, **157**, 275.
55. Gupta,R.C. and Randerath,K. (1979) *Nucleic Acids Res.*, **6**, 3443.
56. Nishimura,S. (1979) In *tRNA: Structure, Properties and Recognition*. Schimmel,P.R., Soll,D. and Abelson,J. (eds), Cold Spring Harbor, USA, p. 551.
57. Tanaka,Y., Dyer,T.A. and Brownlee,G.G. (1980) *Nucleic Acids Res.*, **8**, 1259.

CHAPTER 7

Computing

M.J.BISHOP

1. INTRODUCTION

Both DNA sequencing and computing are in phases of rapid technological change. It is possible to sequence DNA with relatively modest equipment which is inexpensive. The process is highly labour intensive and by investing in instrumentation the labour costs drop but the equipment costs rise steeply. The ultimate aim is a machine which receives a sample of DNA and delivers its sequence on a computer readable medium. This has not yet been achieved though a number of the necessary steps have been automated (1), for example, the GENESIS 2000 DNA Analysis System from Du Pont reads sequencing gels directly into an Apple Mac II computer.

For the majority of laboratories, DNA sequencing results in an autoradiograph of a gel ladder (by the Sanger dideoxy synthetic method or the Maxam−Gilbert degradative method) from which the sequence may be directly read. The day has not yet arrived when flow techniques make the reading of autoradiographs obsolete. The autoradiographs may be entered into a computer manually by typing in each base, digitized by hand, or imaged electronically. Digitizing by hand is the most common method today.

The next stage is the assembly of the cloned fragments into a single, well determined sequence of the DNA. Directed strategies may be best for small lengths of DNA, but for large scale sequencing projects the random strategy is the fastest and most accurate technique. A set of DNA fragments which are related to one another by partial overlap is called a contig. Each DNA fragment in a sequencing project belongs to one and only one contig and each contig contains at least one fragment. The DNA fragments in a contig can be viewed as representing a continuous piece of DNA, the length of which is the length of the contig (*Figure 1*). When a sequencing project is complete there will be only one contig and its consensus will be the entire region of DNA which has been sequenced.

In the random (or shotgun) sequencing technique the DNA to be determined is broken into fragments of about 400 nucleotides in length. These fragments are cloned and the sequences of the cloned inserts are determined. The relationships between the cloned fragments is found by comparing their sequences (2). If one sequence is found to overlap with another then the two fragments can be joined. The process of sequence determination and comparison is continued until the DNA to be sequenced is covered by cloned fragments in a continuous well determined piece. Towards the end of a sequencing project it may be necessary to use a directed strategy to fill gaps between contigs. The DNA will have been sequenced in both directions (on both strands) and a base at a given position will have been determined a number of times so that the possibility of errors is considerably reduced.

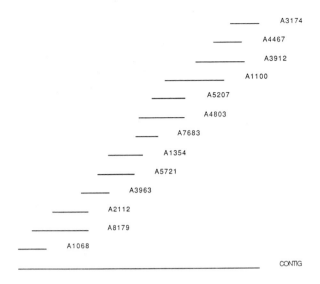

Figure 1. Illustration of the way in which cloned DNA fragments overlap to make up a contig (this is a hypothetical example).

Once the DNA has been sequenced, further computer analysis may be required. Methods have been developed which enable the prediction of the function of unknown DNA. DNA coding for structural RNA shows characteristic inverted repeats (3). Genes coding for protein have characteristic signals for transcription and translation, and the composition of DNA coding for proteins is biased because of the uneven use of amino acids in proteins (4). The functions of the proteins themselves may be indicated by certain sequence motifs or characteristic tertiary structures (5).

The rate at which DNA can be sequenced is becoming sufficiently fast for the development of a new phase in the understanding of cellular development and function at the molecular level. There will soon be adequate data for systematic studies of genome structure, function and evolution. The primary data for such studies are held in computer databases (6). The physical mapping of a genome involves ordering cloned fragments of DNA into contigs. A physical map of the whole *Escherichia coli* genome (4.7 Mb covered by 3400 λ clones) has already been completed (7). Cloned DNA may be maintained in the laboratory to provide the material necessary for functional studies. Genetic map databases, such as the Howard Hughes Medical Institute Gene Mapping Library, provide the top level of the organization of our knowledge of genome structure and function.

Molecular sequence databases comprise a number of both DNA and protein sequence collections made at a number of centres around the world. Submission of a sequence in a machine readable form is becoming a prerequisite for publication in a number of journals. Electronic computer mail expedites this data submission. It is also possible to send an electronic message requesting a sequence from a server at the database centre and the sequence will be automatically returned to the requesting computer. Comparative studies can help give an insight into protein function and evolution. A newly sequenced length of DNA can be screened against all known sequences for identification or for detection of similarities with other sequences (8). There are also databases of macro-molecular structure such as the Protein Data Bank of the Brookhaven National Laboratory. These form the starting point for the prediction of unknown structures from sequence when a sufficiently similar sequence of known structure exists.

1.1 Computer hardware and system software

There is a need for computer methodology in many areas of modern molecular biology and general descriptions are given of the types of hardware and software which are necessary to tackle these tasks. Computers, like means of transport, range from the personal to the supersonic. Every DNA sequencing laboratory needs a personal computer for data entry and may have a more powerful machine for data analysis. Alternatively, a network link to a larger host may be installed and this is useful for communicating with colleagues and with the database centres. The 'supersonic' vector computers and parallel computing arrays are not routinely used in sequencing, although we may expect that parallel computing will become important in the next few years.

Laboratory workstations which can also double as terminals are most usually either an IBM PC (or one of its compatibles) or an Apple Macintosh. Larger machines for a research group are typically the DEC MicroVAX or a Unix workstation (eg. SUN Microsystems). Remote hosts are probably VAXs or Unix systems. This choice of machines brings with it a choice of operating systems and indeed application software and there are four main cultures at the present time: MS DOS, Apple Mac, VAX/VMS and Unix. The MS DOS culture is about to evolve to an OS/2 culture. Price−performance ranges are on the move too. The IBM PC and the Mac machines were in the $1000−5000 range and the VAXs and SUNs around $10000 upwards. Now the IBM PS/2 range includes more powerful machines and there is the Mac 2 which is able to run Unix. There are less expensive versions of MicroVAX and SUN products which are bringing the prices down to meet the upward movement from IBM and Apple.

1.2 Molecular sequence analysis software

The aim is to provide the required functions at minimum cost without sacrificing quality. This is always a matter of local circumstances and, to some extent, of individual preference and the best course is to consult a local expert. Electronic bulletin boards enable one to keep an eye on developments (see Section 1.4). The Journal *CABIOS* and the computer issues of *Nucleic Acids Research* should also be consulted.

Reading a DNA sequencing gel autoradiograph by hand with a digitizing pen requires only a modest computer. An IBM PC with floppy disk drives is adequate. For assembly of contigs and further analysis of the sequence an IBM PC AT (or compatible) or a PS/2 Model 50 with a hard disk is the minimum requirement. There are a number of commercial packages which are suitable (*Table 1*), though these vary considerably in functionality. There are also freely available programs for the IBM PC and many of these are distributed (in the US) by BIONET (address as for IntelliGenetics in *Table 3*) and by MBCRR, Harvard School of Public Health, Dana-Farber Cancer Institute—D1154, 44 Binney Street, Boston, MA 02115, USA, or (in the UK) by The DNA Sequencing Teaching Project, University of Cambridge, Computer Laboratory,

Table 1. Commercial molecular sequence analysis packages for the IBM PC and compatibles.

DNASIS—Pharmacia LKB Biotechnology, Pharmacia House, Midsummer Boulevard, Milton Keynes MK9 3HP, UK.

DNASTAR—DNAStar Ltd, Link House, 565–569 Chiswick High Road, London W4 3AY, UK.

GENEPRO—Riverside Scientific Enterprises, 18332 57th Avenue NE, Seattle, WA 98155, USA.

GENE-MASTER—Bio-Rad Laboratories, Caxton Way, Watford Business Park, Watford, Hertfordshire WD1 8RP, UK.

MICROGENIE—Beckman-RIIC Ltd, Progress Road, Sands Industrial Estate, High Wycombe, Buckinghamshire HP12 4JL, UK.

PC GENE—IntelliGenetics Inc., Amocolaan 2, B-2440 Geel, Belgium.

PUSTELL—IRL Press Ltd, Southfield Road, Eynsham, Oxford OX8 1JJ, UK.

STADEN PLUS—Amersham International PLC, Amersham Laboratories, White Lion Road, Amersham, Buckinghamshire HP7 9LL, UK.

Table 2. Molecular sequence analysis packages for the Apple Macintosh.

Bellon—Bernard Bellon, Laboratoire de Genetique et Biologie Cellulaires CNRS, Faculte de Luminy, Case 907, 13288 Marseille, France.

DNAid—Frederic Dardel, Laboratoire de Biochimie, Ecole Polytechnique, 91128 Palaiseau, France.

DNA Inspector—Textco, 27 Gilson Avenue, West Lebanon, New Hampshire, CT 03784, USA.

DNA Strider—Christian Marck, Departement de Biologie, Centre d'Etudes Nucleaires de Saclay, 9119 Gif-sur-Yvette, France.

MacDNA—Bob Schleif (obtainable from BIONET). Note that this program does not use the Mac mouse and windows environment.

MacGene—from Applied Genetic Technology Inc., 3910 West Valley Drive, Fairview Park, OH 44126, USA.

MacMolly—Soft Gene Berlin, Offenbacher Str. 5, D-1000, Berlin 33, Germany.

Table 3. Molecular sequence analysis packages for VAX/VMS and Unix.

ARP—Hugo Martinez, Department of Biochemistry and Biophysics, School of Medicine, University of California at San Francisco, San Francisco, California, CA 94143, USA.

BION—IntelliGenetics Inc., 1974 El Camino Real West, Mountain View, CA 94040, USA.

EUGENE—Molecular Biology Information Resource, Department of Cell Biology, Baylor College of Medicine, 1, Baylor Plaza, Houston, TX 77030, USA.

GCG—Genetics Computer Group, Biotechnology Center, 1710 University Avenue, Madison, WI 53705, USA.

NBRF—National Biomedical Research Foundation, Georgetown University Medical Center, 3900 Reservoir Road, NW, Washington DC 20007, USA.

PROPHET—Charlotte Hollister, Bolt, Beranek and Newman Laboratories Inc., 10 Moulton Street, Cambridge, MA 02238, USA.

STADEN—Dr. Rodger Staden, MRC Laboratory of Molecular Biology, Hills Road, Cambridge CB2 2QH, UK.

Pembroke Street, Cambridge CB2 3QG. There are also a number of programs for the Apple Macintosh as shown in *Table 2*.

For VAX/VMS popular packages are GCG, IntelliGenetics, NBRF and STADEN (*Table 3*). The NBRF software (NAQ, PSQ) is for indexing and interrogating sequence databases. NBRF also distribute the FASTN and FASTP database searching programs of Lipman, Pearson and Wilbur. For SUN the major packages are ARP, BION, EUGENE and PROPHET. A number of other packages are being moved to Unix (GCG, STADEN) and so there will be more choice in the near future.

1.3 **DNA sequencing software**

Not all the packages mentioned in the previous section contain adequate software for the DNA sequencing process itself. The STADEN package contains the DB System for DNA sequencing. The GCG package has a derivative suite of programs called the Fragment Assembly System. Other adequate tools are the GEL program from IntelliGenetics and the SEQMAN program from DNASTAR. GEL and SEQMAN support Maxam − Gilbert sequencing in addition to Sanger sequencing. An automated film reading and analysis system called Gene-Master is marketed by Bio-Rad (*Table 1*). An automated gel reading system employing fluorescent dye-labelled terminators is marketed by Du Pont under the name GENESIS 2000.

The choice of a suitable computer system for DNA sequencing is of course a matter of personal preference and will also be influenced by the funds available and by the other resources which can be accessed. A person with adequate funding wanting to remain entirely IBM PC based would choose the DNASTAR system while a person wanting the facilities of Unix might choose a SUN workstation running the Intelli-Genetics software. If funding is limited an IBM PC with the Amersham STADEN PLUS package to perform gel entry and contig determination with the ability to connect to a publicly provided VAX/VMS or Unix system for database searching would be optimal.

1.4 **Electronic mail and bulletin boards**

There is an electronic bulletin system for molecular biology called BIOSCI which can be read by logging in to host computers or can be received as electronic mail messages. BIOSCI has a number of different subject boards such as METHODS-AND-REAGENTS and RESEARCH-NEWS (9). BIOSCI is distributed by BIONET on the INTERNET network and requests to receive bulletins in the US should be sent to biosci@net.bio.net. A BIOSCI node on the BITNET network operates from Dublin and requests to receive bulletins should be sent to biosci@irlearn. BIOSCI on the JANET network operates from the SERC Daresbury Laboratory, UK, and requests to receive bulletins should be sent to biosci@uk.ac.daresbury. Unix users may receive BIOSCI bulletins as USENET newsgroups.

2. DATABASES FOR MOLECULAR BIOLOGY

It has been realized for a number of years that there is going to be a crisis in biological information retrieval unless novel approaches are adopted. In the case of nucleotide sequence data it is too difficult to get data into the database merely by having a team to copy and interpret the published literature. It is also the case that the sequence databases have not usually been managed in a Database Management System (DBMS) and that there are therefore inconsistencies. The error rate in transcribing from published work is quite high. It is difficult to interrogate the databases as there is no standard portable software for the purpose. The flat form of the database distribution files was chosen originally to be flexible enough to be processed in any DBMS on any computer system. In reality, the structure of the files is quite rigid and this limits the usefulness of the data.

The whole problem of the interrelation and connection of biomedical databases is under consideration. It has been suggested that the bibliographic database MEDLINE could be connected to GenBank by using the same indexing vocabulary (or at least establishing mappings between the two) (10). The National Library of Medicine has an indexing system called 'Medical Subject Heading' (MeSH) which could be used.

While bibliography is common to all databases, it will be essential to ensure the connection between the other relevant biological databases. Taxonomic databases form the top level of the organization of our knowledge about biological diversity. There are databases of culture collections of microbial strains, cell lines and hybridomas which catalogue sources of materials. There are databases of cloned DNA fragments and it will be important to maintain access to the ordered clones which result from the physical mapping of genomes. Genetic maps resulting from linkage and cytogenetic studies will have to be correlated with the physical map. DNA sequence databases are linked both to the genetic maps which describe the organization of the genetic functions in the genome and to protein sequence databases which specify the translated products. Protein sequence databases are linked to protein structure databases which provide the material for attempts to understand genetic functions at the molecular level. The Human Genome Mapping Projects now being started may provide the necessary momentum to ensure proper coordination of the hierarchy of related databases for molecular biology. A start has been made in so far as there exists a database of databases called 'Listing of Molecular Biology Databases' (LiMB) available from the Los Alamos National Laboratory (request a copy by electronic mail from limb@arpa.lanl).

2.1 DNA sequence databases

There are three major DNA sequence databases:

(i)   the DNA Data Bank of Japan (DDBJ),
(ii)   the European Molecular Biology Laboratory Nucleotide Sequence Data Library (EMBL), and
(iii)   the GenBank Genetic Sequence Data Bank (GenBank).

These use somewhat different data formats although there is collaboration and an effort to standardize. For the current formats consult the manuals distributed with the databases. An example of an entry from the EMBL Data Library is shown in *Figure 2*. It is expected that the databases will be restructured in the near future and that relational tables for the ORACLE DBMS will be distributed allowing SQL (Sequence Query Language) interrogation. The methods of specifying the sequence features will also be improved.

A number of journals (for example, *Nucleic Acids Research*) now require that submitted manuscripts containing sequence data must be accompanied by evidence that the sequences have been deposited at one of the DNA database centres. A submission consists of (i) the sequences in computer readable form and (ii) a completed data submission form for each sequence. The data may be sent by normal post but it is preferable to use electronic mail. The network address for submission to EMBL is datalib@earn.embl and for GenBank is genbank@arpa.lanl.

Improvements in database distribution are being made. The standard method is to distribute the entire database on half inch magnetic tape at quarterly intervals. This has to be read into a computer and the data processed further for the local system. It is now possible to request copies of new (and previous) database entries from the EMBL Network File Server. This is available on BITNET (=EARN) but requests may be made through the ARPA or JANET gateways. We can expect to see distribution of the databases on optical compact disk (CD ROM) becoming increasingly popular. To be used effectively the disks need to be accompanied by software to handle them. LASERGENE from DNASTAR (*Table 1*) is a version of the DNASTAR software on compact disk and permits fast database searches on IBM Personal Computers.

3. READING SANGER GELS

Autoradiographs of Sanger dideoxy sequencing gels may be read by eye, by digitizer or by using a digital scanner (11). As the sequence will have to be entered into a computer for further analysis a computer based method is preferred. For the shotgun sequencing method the gel readings will be assembled by a computer program (2) to yield the final consensus. A number of problems may occur when interpreting gels due either to faulty experimental conditions or sequence-specific interactions (see Chapter 3).

3.1 Staden ambiguity codes

Ambiguity codes which have been specially designed to allow for possible difficulties in interpreting Sanger gels are recognized by the Staden DB System. They are used to allow incompletely determined sequences to be processed by computer. They serve a different purpose from the international uncertainty codes for nucleic acids with which they should not be confused. The Staden codes are shown in *Table 4*.

3.2 Digitizer entry of gel readings

A program called READGEL for Sanger gel reading is available from the DNA Sequencing Teaching Project (Section 1.2) to run on the IBM PC using the SAC GrafBar sonic digitizer. KERMIT (a public domain terminal emulation and file transfer program) may be used to transmit the gel readings to a remote host. READGEL has similar functionality to the program GELIN which is available as part of the Staden DB System on VAX/VMS or Amersham Staden Plus on the IBM PC. The GCG package has a program called GelEnter of similar function for VAX/VMS. The sequencing auto-radiograph is taped to the light box perpendicular to the digitizer. There is a digitizer menu just in front of the digitizer and a program menu taped to the right of the gel. The digitizer menu is activated so that the program can be given the coordinates of the program menu which is then used to control the program. The edges of the set of four lanes to be read are touched with the stylus. Then the sequence is entered into the computer by touching each band in turn, working up the gel (*Figure 3*). A musical tone acts as a check that the correct base has been entered. If mistakes are made it is possible to delete backwards and enter the bands again. The tracks can suffer from distortion either by diverging or by snaking up the gel. It is possible to reset the edges of the track at any stage if the program becomes confused about the correct track positions.

```
ID   HSIL02     standard; RNA; 801 BP.
XX
AC   V00564;
XX
DT   19-AUG-1985  (ref 2 added)
DT   11-APR-1983
XX
DE   Human mRNA encoding interleukin-2 (IL-2)
DE   a lymphozyte regulatory molecule.
XX
KW   interleukin; signal peptide.
XX
OS   Homo sapiens (human, homme, Mensch)
OC   Eukaryota; Metazoa; Chordata; Vertebrata; Tetrapoda; Mammalia;
OC   Eutheria; Primates.
XX
RN   [1]   (bases 1-801; enum. 1 to 801)
RA   Taniguchi T., Matsui H., Fujita T., Takaoka C., Kashima N.,
RA   Yoshimoto R., Hamuro J.;
RT   "Structure and expression of a cloned cDNA for human
RT   interleukin-2";
RL   Nature 302:305-310(1983).
XX
RN   [2]
RA   Devos R., Plaetinck G., Cheroutre H., Simons G., Degrave W.,
RA   Tavernier J., Remaut E., Fiers W.;
RT   "Molecular cloning of human interleukin 2 cDNA and its expression
RT   in E. coli";
RL   Nucl. Acids Res. 11:4307-4323(1983).
XX
```

```
FH   Key             From     To      Description
FH
FT   MSG               1      801     messenger RNA of interleukin-2
FT   CDS              48      506     coding sequence of interleukin-2
FT                                    (48 is 1st base in codon)
FT                                    (506 is 3rd base in codon)
FT   CDS              48      107     coding sequence of signal peptide
FT                                    (48 is 1st base in codon)
FT                                    (107 is 3rd base in codon)
XX
SQ   Sequence 801 BP;   282 A;   147 C;   114 G;   258 U;
     AUCACUCUCU UUAAUCACUA CUCACAGUAA CCUCAACUCC UGCCACAAUG UACAGGAUGC
     AACUCCUGUC UUGCAUUGCA CUAAGUCUUG CACUUGUCAC AAACAGUGCA CCUACUUCAA
     GUUCUACAAA GAAAACACAG CUACAACUGG AGCAUUUACU GCUGGAUUUA CAGAUGAUUU
     UGAAUGGAAU UAAUAAUUAC AAGAAUCCCA AACUCACCAG GAUGCUCACA UUUAAGUUUU
     ACAUGCCCAA GAAGGCCACA GAACUGAAAC AUCUUCAGUG UCUAGAAGAA GAACUCAAAC
     CUCUGGAGGA AGUGCUAAAU UUAGCUCAAA GCAAAAACUU UCACUUAAGA CCCAGGGACU
     UAAUCAGCAA UAUCAACGUA AUAGUUCUGG AACUAAAGGG AUCUGAAACA ACAUUCAUGU
     GUGAAUAUGC UGAUGAGACA GCAACCAUUG UAGAAUUUCU GAACAGAUGG AUUACCUUUU
     GUCAAAGCAU CAUCUCAACA CUAACUUGAU AAUUAAGUGC UUCCCACUUA AAACAUAUCA
     GGCCUUCUAU UUAUUUAAAU AUUUAAAUUU UAUAUUUAUU GUUGAAUGUA UGGUUUGCUA
     CCUAUUGUAA CUAUUAUUCU UAAUCUUAAA ACUAUAAAUA UGGAUCUUUU AUGAUUCUUU
     UUGUAAGCCC UAGGGGCUCU AAAAUGGUUU CACUUAUUUA UCCCAAAAUA UUUAUUAUUA
     UGUUGAAUGU UAAAUAUAGU AUCUAUGUAG AUUGGUUAGU AAAACUAUUU AAUAAAUUUG
     AUAAAUAUAA AAAAAAAAAA C
//
```

Figure 2. A sequence entry from the EMBL Data Library. The meaning of the line codes is as follows: ID, identification; AC, accession number(s); DT, date; DE, description; KW, keywords; OS, organism species; OC, organism classification; RN, reference number; RA, reference author; RT, reference title; RL, reference location; FH, feature table header; FT, feature table data; SQ, sequence header; (blanks), sequence data; XX, spacer line; //, termination line.

193

Table 4. Staden ambiguity codes for Sanger dideoxy DNA sequencing gel interpretation.

Symbol	Meaning	
1	probably C	
2	probably T	
3	probably A	
4	probably G	
D	probably C	possibly CC
V	probably T	possibly TT
B	probably A	possibly AA
H	probably G	possibly GG
K	probably C	possibly C −
L	probably T	possibly T −
M	probably A	possibly A −
N	probably G	possibly G −
R	A or G	
Y	C or T	
5	A or C	
6	G or T	
7	A or T	
8	G or C	
−	A or G or C or T	

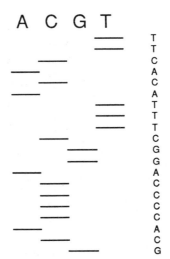

Figure 3. A diagrammatic representation of tracks from a Sanger dideoxy DNA sequencing gel showing how the sequence may be directly read.

4. SCREENING GEL READINGS

Before attempting to find overlaps to build contigs the digitized gel readings and their reverse complements must first be screened for sequences which represent artefacts of the cloning process and are not part of the DNA sequence of interest. The GCG GelEnter program permits screening to be done at the gel reading stage.

```
SCREENV V3.2  AUTHOR: RODGER STADEN
? FILE OF GEL READING NAMES=fof.nam
? FILE NAME FOR PASSED GELS FILE OF FILE NAMES=passed.nam
? VECTOR SEQUENCE FILE NAME=m13mp7.seq
? MINIMUM MATCH (> 11) =12
COMPARING GEL READING dpj10.gel
STRAND   1
STRAND   2
      1  6142

MATCH FOUND WITH CONTIG NUMBER =    1
6122      6132      6142      6152      6162      6172
      CACTCATTAG GCACCCCAGG CTTTACACTT TATGCTTCCG GCTCGTATGT TGTGTGGAAT
      ********** ********** ********** ********** ********** **********
      CACTCATTAG GCACCCCAGG CTTTACACTT TATGCTTCCG GCTCGTATGT TGTGTGGAAT
      1         11        21        31        41        51
6182      6192      6202      6212      6222      6232
      TGTGAGCGGA TAACAATTTC ACACAGGAAA CAGCTATGAC CATGATTACG AATTCCCCGG
      ********** ********** ********** ********** ********** **********
      TGTGAGCGGA TAACAATTTC ACACAGGAAA CAGCTATGAC CATGATTACG AATTCCCCTC
      61        71        81        91        101       111
6242      6252      6262      6272      6282      6292
      ATCCGTCGAC CTGCAGGTCG ACGGATCCGG GGAATTCACT GGCCGTCGTT TTACAACGTC
      **  *  *   **   *      **    *    **   *  *    *  *   * *  * *
      ATTGATGAAG GTGTGATGCC AGATCTTCTC CATATCATCC CAGTTGGTGA TGATACCATG
      121       131       141       151       161       171
6302      6312
      GTGACTGGGA AAACCCTGGC
      * *****  *   * **
      CTCAATGGGG TACTTCAGGG
      181       191
```

Figure 4. A DNA clone has been screened against the vector sequence M13mp7 which is found to be present in the sequence read from the gel. The output is from the program SCREENV.

4.1 **Screening against vector sequences**

It is possible that the gel readings may contain sequences from any of the vectors used during the manipulation of the DNA. Before they are assembled the gel readings must be compared against all the vectors used. The simplest method of doing this is to look for a perfect match of some specified minimum length unlikely to arise by chance (say 15 nucleotides) between the gel reading and the vector. The program in the Staden DB System which does this is called SCREENV. Output from SCREENV is shown in *Figure 4*. Alternatively, an alignment program may be used to display the parts of the gel reading and vector sequence which match. Failure to take the precaution of screening against vector has meant that some published sequences are contaminated.

4.2 **Screening against restriction enzyme sites**

If sonication is used to produce the shotgun fragments for cloning (see Chapter 1, Section 3.3.3), the DNA is first circularized so that the chance of breakage is even along its length. (A linear molecule would be broken more frequently at its ends.) The ligation point is a restriction site and fragments which span this must not be allowed to form overlaps as single sequences. A program capable of finding restriction enzyme recognition sites is needed to detect such gel readings. The program which does this in the Staden DB System is called SCREENR.

5. ASSEMBLING THE CONTIGS

The problem of assembling the gel readings into contigs would be relatively simple if the gel readings were a perfect representation of the DNA sequence. Problems could arise even so if the sequence contains repeats. In practice, the gel readings are imperfect and it is necessary to allow for mismatches and for gaps where bases have been missed. As long as the errors in the gel readings are few an algorithm using perfect matching of short lengths suffices to assemble the contigs. An important principle is that no information should be deleted from the sequencing database by a computer program. Alignments are achieved by putting padding characters into the gel readings and contigs (*Figure 5*).

In a DNA sequencing project, not all the gel readings needed to complete the project are available near the start. The order in which gel readings are added to contigs will cause minor variations in the consensus which is calculated. However, the sequencer will return to re-examine gels which contain problem areas and if necessary will carry out further sequencing, possibly in the complementary strand. For this reason, it is important to file gels systematically at the gel entry stage.

5.1 **Automatic assembly of gel readings**

In the Staden DB System the program which enables automatic assembly of gel readings is called DBAUTO. Both strands of each new gel reading to be added are compared with the current consensus for each contig. (At the start of a new project the current consensus is the first gel added.) Overlaps are found by searching for exact matches of some minimum length (say 15). There are other controls on gel entry.

(i)   The maximum number of pads (gaps) the alignment routine may enter into a contig per gel reading.

```
113              123        133        143        153                    163
CCACAGGGCC AGCTCCAAGA ATGTGAGGGT GTTTCTGGTA CTCTACTACA CATCACAGGC
********** ********** ********** * ********** ********** **********

CCACAGGGCC AGCTCCAAGA ATGTGAGG4T GTTTCTGGTA CTCTACTACA CATCACAGGC
1          11         21         31         41         51
173        183        193        203        213        223

CATCACGGTC ACCTTCATGA GTCAGTGCTT CGCCGGTCGC TGTGGGGCCA ATCAACCGAC
********** ********** ***** **   ********** ********** **********

.CATCACGGTC ACCTTCATGA *TCAGT8*TT CGCCGGTCGC *GTGGGGCCA ATCAACCGAC
61         71         81         91         101        111
233        243        253        263        273        283

CGCCCATTTC TCCATCAGCG TGCCCGCCTC CAGAATCATA AATAGGGCTG AGG
********** ******** * ********** ********** ********** ***

CGCCCATTTC TCCATCA4C4 TGCCCGCCTC CAGAATCATA AATAGGGCTG AGG
121        131        141        151        161        171
```

Figure 5. The program DBAUTO has aligned an existing contig (**above**) with a new gel reading (**below**). Note the ambiguity symbols, mismatches and padding.

197

(ii)   The maximum number of pads the alignment routine may enter into a gel reading.

(iii) The maximum percentage mismatch between the new gel reading and the consensus.

If a new gel reading shows no overlap it is entered into the database to form the start of a new contig. If it shows good overlap with an existing contig, the gel reading and the contig are edited and entered into the database. If a new gel reading overlaps two contigs they will be joined. Sometimes, gel readings are not sufficiently distinct to form the start of a new contig but do not overlap well enough to meet the specified criteria to allow them to be entered into an existing contig by DBAUTO. In this case the program DBCOMP is used to show any overlaps beween the gel reading and existing contigs. DBCOMP is run after the completion of DBAUTO because further gels processed since the problem gel could have altered the consensus.

5.2 Displaying the contents of a database

The program which allows display of the contents of a database and the editing of contigs is called DBUTIL. There is an option to print the relational information for contigs (*Figure 6*) which can be converted into a diagram showing the overlaps (*Figure 7*). Another option displays the aligned sequences in a contig (*Figure 8*). There are also facilities to edit gel readings, enter new gel readings manually, complement and join contigs and alter the relationship of gel readings.

6. ASSESSING THE QUALITY OF SEQUENCED DNA

The object of a DNA sequencing project is to produce a consensus sequence from all the aligned gel readings. The quality of a consensus depends on the number of times it has been sequenced and on the uncertainties used in each gel reading. A program HIGH is provided to highlight areas where further attention is needed (*Figure 9*). There are also options of DBUTIL for examining the quality in quantitative terms. When a project is nearing completion it is necessary to clean up whole contigs and a screen editing procedure is provided to do this. If it is suspected that some overlaps have been missed by the automatic assembly procedure the program ENDSOUT can be used to write out the gel readings at the ends of contigs. DBCOMP can then be used to compare these with the consensus for the whole database. If the gel readings match more than one contig this indicates a possible join that can be made.

The final product of the sequencing project is a consensus sequence which does not exist in the database but which is calculated from the database whenever needed. Individual characters are assigned the following scores:

definite assignments A,C,G,T,B,D,H,V,K,L,M,N = 1
probable assignments 1,2,3,4 = 0.75
any other character = 0

For each position in a contig four base totals are calculated for A,C,G,T respectively. These totals are calculated by adding up the individual scores for each of the characters for a base at a position e.g. A,B,M,3 contribute to the total for A. If the total for one base is greater than or equal to 3/4 of the sum for that position it is included in the consensus. Otherwise a '−' character is assigned to indicate uncertainty.

CONTIG	LINE	LENGTH		ENDS	
				LEFT	RIGHT
	45	426		7	22
NAME	NUMBER	POSITION	LENGTH	NEIGHBOURS	
				LEFT	RIGHT
MBE122.;1	7	1	−196	0	6
MBE114.;1	6	180	105	7	22
MBE161.;1	22	197	−230	6	0

CONTIG	LINE	LENGTH		ENDS	
				LEFT	RIGHT
	46	95		5	5
NAME	NUMBER	POSITION	LENGTH	NEIGHBOURS	
				LEFT	RIGHT
MBE110.;1	5	1	95	0	0

CONTIG	LINE	LENGTH		ENDS	
				LEFT	RIGHT
	47	333		4	3
NAME	NUMBER	POSITION	LENGTH	NEIGHBOURS	
				LEFT	RIGHT
MBE104.;1	4	1	221	0	11
MBE134.;1	11	160	167	4	3
MBE100.;1	3	179	155	11	0

CONTIG	LINE	LENGTH		ENDS	
				LEFT	RIGHT
	48	153		2	2
NAME	NUMBER	POSITION	LENGTH	NEIGHBOURS	
				LEFT	RIGHT
MBE098.;1	2	1	153	0	0

CONTIG	LINE	LENGTH		ENDS	
				LEFT	RIGHT
	49	433		1	12
NAME	NUMBER	POSITION	LENGTH	NEIGHBOURS	
				LEFT	RIGHT
MBE025.;1	1	1	285	0	8
MBE126.;1	8	113	−297	1	12
MBE136.;1	12	243	191	8	0

Figure 6. This output is from the 'Show Relationships' option of DBUTIL and shows the state of a DNA sequencing database near the start of a new project. There are five contigs, three of which contain three gel readings and two of which consist of only one gel reading.

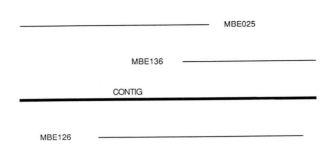

Figure 7. This representation of a contig shows each gel reading as a line proportional to the length of the sequence. The two strands are shown above and below the solid line representing the contig.

7. COMPUTER ANALYSIS OF SEQUENCES

7.1 Functional types of nucleic acid sequences

DNA codes for structural RNAs (transfer RNAs, ribosomal RNAs) and proteins, including the signals for transcription, translation and protein function. The function of RNAs and proteins often depends on macromolecular interactions in three dimensions and the majority of these interactions are poorly or not understood at present. Computer methods can help in the task of assigning functions to DNA sequences but the ultimate verification must come from experiments in which the products of transcription and translation are characterized and their interactions with other molecules are studied experimentally.

In eukaryotes as little as 1% or less of the DNA may code for structural RNA and protein. In addition, there are many families of repetitive sequences the function of which is not understood. Repeated sequences may be involved in the organization of chromosomes and in recombination. Some genes, such as those for rRNAs may be present in many copies or be present in multigene families with many related members which appear to have arisen by gene duplication. Other genes, such as those for the proteins chorion and collagen, may contain many repetitive elements within the coding sequence.

The mRNAs which are transcribed from DNA may (in eukaryotes) undergo complex processing. Segments of sequence called introns may be excised between the protein coding exons. Pseudogenes also exist, which are derelict genes that remain in the DNA after they have ceased to function and their signals or coding regions are degraded. Some pseudogenes appear to have been derived by reverse transcription from mRNA as they contain no intron sequences.

Coding for protein has an important effect on the composition of DNA as described in Section 7.2. Signals for transcription and translation may be located by methods described by Stormo (4). The combination of DNA composition and signals can be quite effective in locating genes coding for protein (12). Coding for structural RNA will impose symmetry properties such as inverted repeats which enable RNA to fold into its secondary structure. Perfect repeats can be found by the methods described in Section 7.3 and methods to predict RNA secondary structure are described by Gouy (3). Computer methods will be invaluable in the prediction of the function of DNA as large regions of the genome are sequenced.

```
                       10        20        30        40        50
 1  MBE025.;1  TGCCAGAAGACAAATTCCCCCAGTTTCTCCATGTCACGTTTGAGCAAACC
    CONSENSUS  TGCCAGAAGACAAATTCCCCCAGTTTCTCCATGTCACGTTTGAGCAAACC

                       60        70        80        90       100
 1  MBE025.;1  GGGCCAAACTTAGTTCAGGTGTGCCATGCCCGGGGCAGGAACTTTGCGTG
    CONSENSUS  GGGCCAAACTTAGTTCAGGTGTGCCATGCCCGGGGCAGGAACTTTGCGTG

                      110       120       130       140       150
 1  MBE025.;1  CCTGAG88ATACCCACAGGGCCAGCTCCAAGAATGTGAGGGTGTTTCTGG
-8  MBE126.;1            CCACAGGGCCAGCTCCAAGAATGTGAGG4TGTTTCTGG
    CONSENSUS  CCTGAG--ATACCCACAGGGCCAGCTCCAAGAATGTGAGGGTGTTTCTGG

                      160       170       180       190       200
 1  MBE025.;1  TACTCTACTACACATCACAGGCCATCACGGTCACCTTCATGAGTCAGTGC
-8  MBE126.;1  TACTCTACTACACATCACAGGCCATCACGGTCACCTTCATGA*TCAGT8*
    CONSENSUS  TACTCTACTACACATCACAGGCCATCACGGTCACCTTCATGAGTCAGTGC

                      210       220       230       240       250
 1  MBE025.;1  TT1GCCGGTCGCTGTGGGGCCAATCAACCGACCGCCCATTTCTCCATCAG
-8  MBE126.;1  TTCGCCGGTCGC*GTGGGGCCAATCAACCGACCGCCCATTTCTCCATCA4
12  MBE136.;1                                            TCCATCAG
    CONSENSUS  TTCGCCGGTCGCTGTGGGGCCAATCAACCGACCGCCCATTTCTCCATCAG

                      260       270       280       290       300
 1  MBE025.;1  CGTG1CCGCCT1CAGAATCATAAATAGGGCTGAGG
-8  MBE126.;1  C4TGCCCGCCTCCAGAATCATAAATAGGGCTGAGGCCAGTCAAGACAGCA
12  MBE136.;1  CGTGCCCGCCTCCAGAATCATAAATAGGGCTGAGG*DAGTCAAGACAGCA
    CONSENSUS  CGTGCCCGCCTCCAGAATCATAAATAGGGCTGAGGCCAGTCAAGACAGCA

                      310       320       330       340       350
-8  MBE126.;1  CTACATCCCAGCTAGCCCGTCGTAGAGACAGACAAGATGGTTCCTTCTCA
12  MBE136.;1  CTACATCCCAGCTAGCCCGTCGTAGAGACAGACAAGATGGTTCCTTCTCA
    CONSENSUS  CTACATCCCAGCTAGCCCGTCGTAGAGACAGACAAGATGGTTCCTTCTCA

                      360       370       380       390       400
-8  MBE126.;1  GAGACTCTCCCGAACTAGCAGCATTTCCTCCAACGAGGATCCCGCTGGTA
12  MBE136.;1  GAGACTCTCCCGAACTAGCAGC*TTTCDTCCAAGCAGGATCCCGCAGGTA
    CONSENSUS  GAGACTCTCCCGAACTAGCAGCATTTCCTCCAA--AGGATCCCGC-GGTA

                      410       420       430       440       450
-8  MBE126.;1  AGAAGCTAC
12  MBE136.;1  AGAAGCTACACCGGCCAGTGG1C4GGCCGTGGA
    CONSENSUS  AGAAGCTACACCGGCCAGTGGCCGGGCCGTGGA
```

Figure 8. The 'Display' option of DBUTIL shows the details of the overlap of three gel readings forming a contig.

7.2 Nucleotide sequence composition

In contemplating the long linear sequence of symbols representing DNA a great many statistical models have been set up to study sequence properties and variation. A relatively limited number of these have found practical application. Nucleotides may be counted

```
                    10        20        30        40        50
  1   MBE025.;1   ..........................................................
                  TGCCAGAAGACAAATTCCCCCAGTTTCTCCATGTCACGTTTGAGCAAACC

                    60        70        80        90       100
  1   MBE025.;1   ..........................................................
                  GGGCCAAACTTAGTTCAGGTGTGCCATGCCCGGGGCAGGAACTTTGCGTG

                   110       120       130       140       150
  1   MBE025.;1   ......88..................................................
 -8   MBE126.;1                ..............................4.........
                  CCTGAG--ATACCCACAGGGCCAGCTCCAAGAATGTGAGGGTGTTTCTGG

                   160       170       180       190       200
  1   MBE025.;1   ..........................................................
 -8   MBE126.;1   .............................................*.....8*
                  TACTCTACTACACATCACAGGCCATCACGGTCACCTTCATGAGTCAGTGC

                   210       220       230       240       250
  1   MBE025.;1   ..1.......................................................
 -8   MBE126.;1   ..........*...............................4
 12   MBE136.;1                                              .......
                  TTCGCCGGTCGCTGTGGGGCCAATCAACCGACCGCCCATTTCTCCATCAG

                   260       270       280       290       300
  1   MBE025.;1   ....1......1.....................
 -8   MBE126.;1   .4........................................................
 12   MBE136.;1   ...............................*D.............
                  CGTGCCCGCCTCCAGAATCATAAATAGGGCTGAGGCCAGTCAAGACAGCA

                   310       320       330       340       350
 -8   MBE126.;1   ..........................................................
 12   MBE136.;1   ..........................................................
                  CTACATCCCAGCTAGCCCGTCGTAGAGACAGACAAGATGGTTCCTTCTCA

                   360       370       380       390       400
 -8   MBE126.;1   ..................................CG..........T....
 12   MBE136.;1   .......................*....D.....GC..........A....
                  GAGACTCTCCCGAACTAGCAGCATTTCCTCCAA--AGGATCCCGC-GGTA

                   410       420       430       440       450
 -8   MBE126.;1   ........                      ................
 12   MBE136.;1   ....................1.4.....................
                  AGAAGCTACACCGGCCAGTGGCCGGGCCGTGGA
```

Figure 9. The program HIGH highlights the problem areas in a contig so that the sequencer may refer to the original gels to check problems or may decide to sequence further DNA to resolve the difficulties.

in singlets, in doublets, in triplets or in longer oligomers. These counts can be considered as being overlapping or non-overlapping and in the non-overlapping cases there are different phases for starting the counts (*Figure 10*). Chosen sections of the sequence may be compared with others or with the entire sequence.

Rather than considering regions of the sequence chosen by some functional criterion

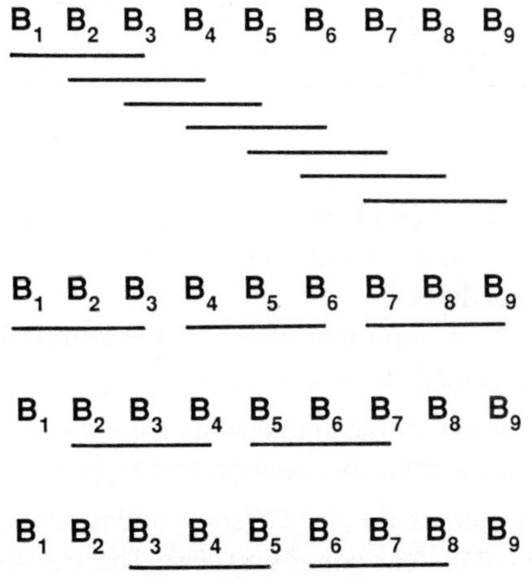

$$B_1 \quad B_2 \quad B_3 \quad B_4 \quad B_5 \quad B_6 \quad B_7 \quad B_8 \quad B_9$$

$$B_1 \quad B_2 \quad B_3 \quad B_4 \quad B_5 \quad B_6 \quad B_7 \quad B_8 \quad B_9$$

$$B_1 \quad B_2 \quad B_3 \quad B_4 \quad B_5 \quad B_6 \quad B_7 \quad B_8 \quad B_9$$

$$B_1 \quad B_2 \quad B_3 \quad B_4 \quad B_5 \quad B_6 \quad B_7 \quad B_8 \quad B_9$$

Figure 10. Molecular sequence analysis of triplets, **above**, overlapping, **below**, in three phases.

(e.g. a coding region) a graphical representation of changes along the whole sequence can be calculated in a sliding window. The window is of fixed length and from the nucleotides in the window the statistic is calculated. The window is repositioned a fixed number of residues further down the sequence and the statistic is recalculated. The values of the statistic are plotted to form a graph.

In all cases suitable statistics will be computed according to the model chosen. Further computations of hypothetical values given certain criteria may also be required. In cases where hypothetical values are hard or impossible to obtain analytically it may be necessary to resort to randomization tests. That is, the sequence symbols are shuffled and the statistic is computed, the process being repeated many times. A sensible and down to earth account of many aspects of sequence analysis (especially proteins) including 'jumbling tests' has been given by Doolittle (13).

7.2.1 *Nucleotide composition*

By moving along a sequence a base at a time it is easy to count the frequency of each base. From this the composition of the complementary strand can be derived. The composition of regions of a sequence may be compared with each other and with regions of the sequence as a whole. This may be done for a single base or for any combination. Such comparisons are not usually very informative. It has been noted that the DNA of warm-blooded animals has a high $G+C$ content which correlates with the higher thermal stability of GC over AT base pairs.

7.2.2 *Dinucleotide composition*

Dinucleotide composition can be measured either as overlapping doublets or as non-overlapping doublets in two phases in one strand and two phases in the other. It is found that CG dinucleotides are less abundant than expected in vertebrates and it is thought that this is because of the problem of repairing methylated CpG because of spontaneous deamination of 5-methylcytosine to thymine (14).

7.2.3 *Trinucleotide composition*

Trinucleotide composition can be measured either as overlapping triplets or as non-overlapping triplets in three phases in one strand and three phases in the other. In regions coding for protein one (or more) of these phases will be the phase of translation and the triplets will be codons.

Coding for protein has an important effect on the composition of DNA and there are three factors which contribute to this (15).

(i) *Uneven use of amino acids*. Some amino acids are used more frequently than others in proteins. For example, alanine is common and tryptophan is rare.

(ii) *Uneven numbers of codons for the different amino acids*. There is considerable variation in the numbers of codons per amino acid. There is minor variation according to the genetic code being used. For example, in the mammalian mitochondrial code arginine is represented by four codons and methionine and tryptophan by two codons.

(iii) *Uneven use of codons*. In a gene the possible codons for each amino acid will not be evenly used.

There are two important aspects to the study of codon frequencies: (i) the study of codon preferences when the position of protein coding sequences are known, and (ii) the prediction of the position of genes in DNA sequences coding for as yet unidentified proteins.

7.2.4 *Codon preferences*

All organisms which have been studied show a codon preference when the usage of codons is compared with the expected random usage for DNA of the same composition (16). It has been suggested that abundance of the different tRNA molecules corresponding to the codons is used to control gene expression (17). Regulatory genes of *E.coli* have higher numbers of rare codons than non-regulatory genes (18). However, mammalian mitochondrial genes show strong codon preferences even when there is only one possible tRNA per amino acid (19).

The frequency of occurrence of one member of a pair of complementary codons is significantly positively correlated with the frequency of occurrence of the other member of the pair. This has the result that the non-coding strand, read in the same phase and with the reverse polarity, tends to resemble the coding strand in codon frequency (20).

7.2.5 *Predicting the position of protein coding genes*

Search for protein coding genes may be made by investigating DNA composition or by search for signals of transcription and translation (4). The most obvious constraint on DNA composition is the absence of stop codons since proteins are generally over 100 amino acids in length. However, in eukaryotic DNA the case is complicated by the presence of introns. Most exons are at least 20 codons long and usually at least 50 codons long. In genes with high G+C content the expected random frequency of

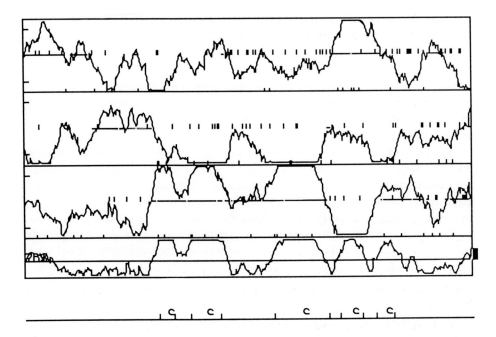

Figure 11. Prediction of protein coding regions in human cytoplasmic β-actin (EMBL database entry HSACCYBB, 3646 bp) using the Staden program Analyseq. The top three curves are for the positional base preference method, ticks along the middle are stop codons, ticks along the bottom are methionines (potential starts). The lowest curve is for the uneven positional base frequency method. The map shows the known position of exons (marked *c*).

stop codons is decreased. Coding constraints affect both strands of the DNA and coding regions tend to have long open reading frames (ORFs) even on the non-coding strand.

(i) *Uneven positional base frequencies*. The observed frequencies in a window are counts of the number of each base at the first, second and third positions along the sequence. The expected frequencies are, for each base, its proportion in the sequence times the window length divided by three. This method cannot indicate the correct reading frame or strand of the DNA. It makes no assumptions about amino acid composition or codon preferences (*Figure 11*).

(ii) *Positional base preferences*. This method uses the fact that protein sequences have an average amino acid composition that generates positional base preferences which are sufficiently characteristic to choose between reading frames. The observed frequencies in a window are calculated for each frame for each codon position [1,2,3]. The expected frequencies are calculated from the average amino acid composition assuming no codon preference (*Figure 11*).

(iii) *Codon preference*. All organisms deviate from a random choice of synonymous codons. The known codon usage for a given species can be used to identify further unknown coding regions in that species. The method examines the fraction of a codon being used for its amino acid in a window and compares this with the known codon usage figures.

7.3 **Repeated nucleotide sequences**

While coding for protein tends to impose certain constraints on DNA composition (Section 7.2), coding for RNA molecules will tend to impose symmetry properties such as inverted repeats which enable the RNA to fold into its secondary structure (3). Repeats and other features with symmetry may be either perfect where there is an exact match of each nucleotide or imperfect where only a proportion of nucleotides match.

7.3.1 *Sequence dictionary*

A sequence dictionary is a very effective way of studying perfect repeats in sequences. The dictionary is built by considering each overlapping oligonucleotide of a certain length along the sequence and sorting these into lexographical order. There are efficient algorithms for doing this (21). The positions of exactly matching oligonucleotides up to a certain length can easily be read from the dictionary (*Figure 12*).

The expected length of the longest direct repeat in a sequence has been given by Karlin *et al.* (22). Consider a randomly generated sequence of length n based on an r letter alphabet with probability P_i of getting one of the r letters. Let $\lambda = \sum_{i=1}^{r} P_i^2$. The length of the expected largest direct repeat is $2\log n/\log (1/\lambda) - [1 + \log (1 - \lambda)/\log \lambda + 0.5772/\log \lambda] + \log 2/\log \lambda$. The variance is $1.645(-1/\log \lambda)^2$ independent of n.

Inverted repeats (sometimes called dyad symmetries) can also be found by the dictionary method by including the DNA strand and its reverse complement. It is then necessary to look up exact matches between one strand and the other (*Figure 13*).

7.3.2 *The dot-plot*

Another way to look for repeats and inverted repeats in sequences is the 'dot-plot'. The dot-plot is a diagram which is a rectangular array with rows labelled by one copy of the sequence and columns labelled by another copy (or the reverse complement to detect inverted repeats). A cell i,j can be used to represent the result of comparison of the jth residue of the first sequence (A) with the ith residue of the second (B) (*Figure 14*). The simplest kind of dot-plot is created by putting a diagonal mark in each cell where symbols match in the two copies of the sequence. The main diagonal of the diagram becomes a solid line where all symbols match and the diagram is symmetrical about this diagonal so that only half need be considered. For nucleic acids where there are only four symbols the diagram will become very cluttered with unwanted dots. The positions of repeats greater than a certain length can be clarified by drawing only diagonals which reach a threshold length. Partial matches may be displayed by choosing a diagonal span of fixed length (for example, 21) and drawing diagonals which reach a certain score (for example, 17) in this span (*Figure 15*).

7.4 **Mapping nucleotide sequence features**

Mapping the position of features on a DNA sequence is an important step in investigating its structure and function. Mapping of sites which can be precisely defined (such as restriction enzyme recognition sites or stop codons) is easy. Mapping other features such as promoters, splice junctions, and ribosome binding sites is less reliable because they cannot be precisely specified.

```
AAAGTTAGGTAAAGTTAGGTAA
TBREP1    +           313
TBREP1    +           323
TBREP1    +           333
AAGTTAGGTAAAGTTAGGTAAA
TBREP1    +           314
TBREP1    +           324
TBREP1    +           334
AGGTAAAGTTAGGTAAAGTTAG
TBREP1    +           309
TBREP1    +           319
TBREP1    +           329
TBREP1    +           339
AGTTAGGTAAAGTTAGGTAAAG
TBREP1    +           315
TBREP1    +           325
TBREP1    +           335
GGTAAAGTTAGGTAAAGTTAGG
TBREP1    +           310
TBREP1    +           320
TBREP1    +           330
GTAAAGTTAGGTAAAGTTAGGT
TBREP1    +           311
TBREP1    +           321
TBREP1    +           331
GTTAGGTAAAGTTAGGTAAAGT
TBREP1    +           316
TBREP1    +           326
TBREP1    +           336
TAAAGTTAGGTAAAGTTAGGTA
TBREP1    +           312
TBREP1    +           322
TBREP1    +           332
TAGGTAAAGTTAGGTAAAGTTA
TBREP1    +           308
TBREP1    +           318
TBREP1    +           328
TBREP1    +           338
TTAGGTAAAGTTAGGTAAAGTT
TBREP1    +           317
TBREP1    +           327
TBREP1    +           337
Repeats found at oligonucleotide length 22
```

Figure 12. Direct repeats of length 22 in the sequence of a *Trypanosoma brucei* kinetoplast DNA minicircle (EMBL database entry TBREP1, 1004 bp).

```
AATATATT
HSUG2    +           276
HSUG2    -            97
ATATTAAA
HSUG2    -           129
HSUG2    +           279
GGTGCACC
HSUG2    +           371
HSUG2    -             2
TTTAATAT
HSUG2    -            94
HSUG2    +           244
Longest repeats found at word length   8
```

Figure 13. Inverted repeats of length 8 in the sequence of the human small nuclear RNA U2 24A (EMBL entry HSUG2, 379 bp).

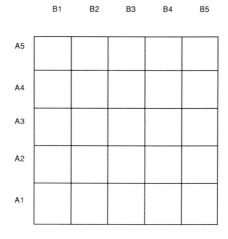

Figure 14. The array for creating a dot-plot between sequences A and B numbered from bottom left to top right as in the program Diagon.

7.4.1 *Expected spacing of recognition sites*

If we assume that recognition sites occur randomly along DNA, then for long strands where the occurrence of individual base pairs is independent, the spacing of recognition sites depends on the individual recognition sequence and the composition of the DNA in terms of its GC/AT content. Let f_{GC} be the fraction of $G+C$ and f_{AT} be the fraction of $A+T$ so that $f_{GC} + f_{AT} = 1$. For a single site the probability P_X of finding base X (X = A,C,G,T) is

$$P_G = P_C = f_{GC}/2$$
$$P_A = P_T = f_{AT}/2$$

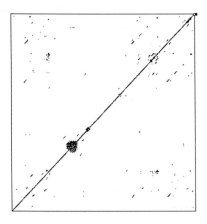

Figure 15. Dot plot of *Trypanosoma brucei* kinetoplast DNA minicircle against itself drawn by the Staden program Diagon using a span of 21 and a threshold score of 17. The strong diagonal line represents identity of the sequence. Repeated regions are off-diagonal and in mirror image.

the factor 2 taking into account the fact that the DNA is double stranded with occurrence on either strand equiprobable but mutually exclusive. Assuming independence of base occurrence the probability P of any given sequence of bases can be written as the product of independent probabilities. For a restriction site of m_{GC} base pairs and n_{AT} base pairs

$$P = (f_{GC}/2)^{m_{GC}} \cdot (f_{AT}/2)^{n_{AT}}$$

and the expected spacing is $1/P$ (23). Deviations from expected behaviour can be used to detect repeated sequences or methylation. For example, the 'Alu family' of repeats (24) was first detected because of behaviour on cleavage with *Alu*I. The enzyme *Hpa*II has enabled the discovery of *Hpa*II tiny fragment (HTF) islands in the mammalian genome (25). There are about 30000 of these $1-2$ kb regions rich in G+C and non-methylated at CpG. Many appear to be associated with genes and in particular with the regions where transcription begins.

7.4.2 *Mapping restriction enzyme recognition sites*

Restriction enzyme recognition sites can be precisely specified and it is therefore a simple matter to locate them in a DNA sequence. If it is required to find the sites for one enzyme it is best to read through the DNA once, noting the positions of the sites. If a map for many enzymes is wanted then it is quicker to make a dictionary from the sequence and look up the positions for each enzyme in turn. There are two useful representations of the results of the analysis. One is a map of the DNA with the positions of the sites marked. The other is a sorted list of fragment sizes which can be compared with the results of gel electrophoresis. Tables of restriction enzymes with their recognition sites and cut points are given by Roberts (26). The following facts about restriction enzymes should be noted.

(i) *Palindromes*. A palindrome is a DNA sequence which reads the same as its reverse complement. Many restriction enzyme recognition sites are palindromes. (Note: this

Table 5. Codes for nucleotides including uncertainties.

G = Guanine
A = Adenine
T = Thymine
C = Cytosine
R = Purine (A or G)
Y = Pyrimidine (C or T or U)
M = Amino (A or C)
K = Ketone (G or T)
S = Strong interaction (C or G)
W = Weak interaction (A or T)
H = Not-G (A or C or T) H follows G in the alphabet
B = Not-A (C or G or T) B follows A
V = Not-T (not-U) (A or C or G) V follows U
D = Not-C (A or G or T) D follows C
N = Any (A or C or G or T)
U = Uracil (in RNA)

These codes are the IUPAC-IUB Commission recommendations and are expected to be adopted as a standard.

differs from the usual meaning of palindrome which is a word which reads the same in either direction.)

(ii) *Incompletely specified bases.* Some restriction enzymes recognize sites which are not precisely specified in terms of A, C, G, and T. They may be precisely specified in terms of the nucleic acid uncertainty symbols (*Table 5*). For example the enzyme *Hinc*II recognizes GTTAAC, GTTGAC, GTCAAC, and GTCGAC and so may be specified by GTYRAC.

(iii) *Asymmetric recognition sites.* Enzymes which recognize asymmetric sites in double-stranded DNA have two different recognition sites. For example, the enzyme *Mbo*II recognizes GAAGA on one strand and TCTTC on the other so that a search for both must be made.

(iv) *Cutting sites.* A restriction enzyme will not necessarily cut DNA at the recognition site but may cut some distance away. An example is *Mbo*II which cuts 13 bases after the site GAAGA and 7 bases before the site TCTTC. A restriction enzyme database contains a list of both recognition sites and cutting sites. Programs must take cutting sites into account when calculating fragment sizes.

(v) *Isoschizomers.* This is the name given to restriction enzymes with the same recognition site, for example *Mbo*I and *Hac*I both recognize GATC. They do not necessarily have the same cutting site.

(vi) *Methylation.* Methylation within the recognition site may affect the activity of an enzyme at that site. Individual isoschizomers may be affected differently.

7.5 Sequence comparison

It is straightforward to compare sequences which are identical. As soon as sequences differ there is considerable scope for argument about how different (or similar) they are. Biologists tend to feel that there must be some unique optimal statistical measure of sequence similarity ('homology'), yet an infinite number of statistical models for sequence comparison can be invented. Therefore the goal should be to find measures of sequence similarity which correlate with biologically significant properties while at the same time being statistically respectable. The purpose for which the comparison is being made and the assumptions one is prepared to make are important in determining the appropriate statistical model and the computer method. If the sequence of a piece of DNA coding for an unknown protein has been determined it may be sufficient for identification to compare a short peptide from the protein for identity with a sequence in a protein database. Very distantly related proteins (whether related by evolutionary descent or by common function) will not be detected by this method and an exhaustive database search may be the only way to reveal anything.

7.5.1 *Searching sequence databases*

Sequence database searching programs can either be exhaustive or can take short cuts in an attempt to reduce the time they take. The time taken to compare two sequences is proportional to the product of their lengths whereas the time taken to locate identities between sequences is proportional to the sum of their lengths so the savings can be dramatic. The most popular short cut programs are the 'FAST' suite: FASTN (27), FASTP (28) and FASTA (29). These gain much of their speed by computing a table to locate identities of up to a fixed threshold length between the probe and the database sequence in the first step. The best 'diagonal' regions (so called because they would show on a dot-plot, *Figure 16*) are then scored by a proportional algorithm. An attempt is made to join the regions so found. If a satisfactory score is obtained an optimal alignment is then calculated.

There is a minor risk with short cut methods that something interesting will be missed. In exhaustive methods the best location for the match of a probe sequence within a longer database sequence or the best local similarity within probe and database sequence is determined by alignment algorithms which find the best path through a graph (Section 7.5.2). These methods are impractical on conventional computers but become attractive when run on computers of parallel architecture (8).

7.5.2 *Sequence alignment*

Sequence alignments are computed using a score matrix based on a graph which is like that used for the construction of dot-plots (*Figure 14*). In any cell, horizontal moves represent a deletion in sequence A, vertical moves represent a deletion in sequence B, and diagonal moves from bottom left to top right represent an identity or a substitution. A path through the graph can be converted into a sequence alignment (30). All possible moves are scored and the task is then to analyse the graph in one of three ways.

(i) *Total alignment*. This is the alignment of the whole of sequence A with the whole of sequence B (*Figure 17*).

(ii) *Best location*. This is the alignment of sequence A (a short sequence) at the best location within a longer sequence B. Every location in B is a potential starting point for the alignment.

(iii) *Best local similarity*. This is the best alignment of a part of sequence A with a

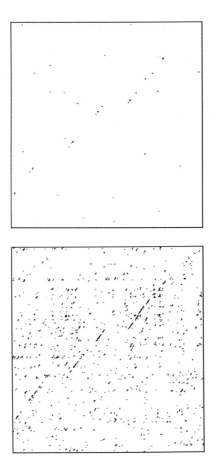

Figure 16. Dot-plot comparison of (**vertical**) human α-globin germline DNA (EMBL database entry HSAGL1, 1138 bp) with (**horizontal**) a mouse pseudogene for α-globin (EMBL database entry MMAGLX, 900 bp) lacking intervening sequences. Breaks in the principal diagonal mark the position of introns. **Above**: identities only; **below**: proportional matches illustrating the first two steps in the program FASTN.

part of sequence B. Every location in sequences A and B are potential starting points for the alignment.

It should be noted that for any arbitrary sequences one or more optimal alignments can be found. Alignments are not data but hypotheses of relationship under the chosen scoring scheme. If an evolutionary view is taken then an alignment implies a series of events which have occurred during the divergence of the two sequences.

It is often of interest to construct a multiple alignment for a family of related sequences. Instead of finding paths through a graph enclosed by a rectangle (*Figure 14*) it is necessary to investigate paths enclosed by a hypercube of dimensions equal to the number of sequences. This is a horrendous task and has not been attempted for more than a few sequences. Rather than computing a joint multiple alignment on all the sequences simultaneously it is possible to construct a multiple alignment from alignments of pairs of sequences in a stepwise fashion (31,32). One strategy (32) is to score all possible

```
CCTAACTTCTTCCCAAACACCAGATGTTGGAGGCAGAGGAATGAGCAAGAAGCATCCGGGCAACTGA

   CC       GC G         CC    C C      GG         G CT  C     C   G
   ::       :: :         ::      : :    ::         : ::  :     :   :
CATCCAAGGAAGGCAGATCCCCAACCCAGGACTCTTGAGGGCTCTTAAAAGCCTGGCTGGGCAGAGC

TGAGCGCCGCCCGGCCGG GCGTGCCCCCGCGCCCCAAGCATA AACCCTGGCGCGCTCGCGGCCCG
::::  :: : :       :: :: ::    : : ::      : ::::: ::      ::  :    :: ::
TGAGGACCACTCAAAGGGCGTGTCTCAAAGAGCTGTAGGCATAGAAGTGCTGCTTATTTCAGGTCCA

CCCCACAGACTCAGAGAGAACCCACCATGGTGCTGTCTCCTGCCGACAAGACCAACGTCAAGGCCGC
 :  :  :::::::  ::  :::  :::::::::::  :::  ::: ::  :::: : :    ::      ::
TCTGATAGACTC  AG GAAGACACCATGGTGC  TCT CTG CG GAAGA C A    AA      GC

GGCGCGCACGCTGGCGAGTATGGTGCGGAGGCCCTGGAGAGGTGAGGCTCCCTCCCCTGCTCCGACC
 :  :    : :  ::     :  :  :  :::  ::   ::    :    :::: ::
GTC AG A ACT    G  T CTG GGA    C  AAGA GT  GC AGGT  CGTGCT   GA

CGCCCGGACCCACAGGCCACCCTCAACCGTCCTGGCCCCGGACCCAAACCCCACCCCTCACTCTGCT
 :: :::   : : ::::    ::            ::         ::    :    :        :
GCGCGG   A A GCCA    AA            GG      A      A      GGA      TG

TGTTCCTGTCCTTCCCCACCACCAAGACCTACTTCCCGCACTTCGACCTGAGCCACGGCTCTGCCCA
 :  : ::  :   : :  ::  :::::::::::::::::::   :::: ::::: ::  :  :  : ::
G    C AT C T  CC CCACCAAGACCTACATCCCTCACTTTGA    T  G TA  G     G CC

GGCAAGAAGGTGGCCGACGCGCTGACCAACGCCGTGGCGCACGTGGACGACATGCCCAACGCGCTGT
 ::              :: :  : : :  ::::  :  ::      :   :  : ::  : :  :: :
GC              CCCCA G G T  CCAA G  GT      A AT    TG  ATG  C A  CG  G

CCTGCACGCGCACAAGCTTCGGGTGGACCCGGTCAACTTCAAGGTGAGCGGCGGGCCGGGAGCGATC
 ::   :  : ::  :: :     : :  :: :  :  : :       :::           : ::       : ::
TG   C TGCA AA C  C   ATGG   CCTG C   C T       GGT           GCC       CTAT

GAGATGGCGCCTTCCTCTCAGGGCAGAGGATCACGCGGGTTGCGGGAGGTGTAGCGCAGGCGGCGGC
 : :      :: :  : :::      :        : ::  ::      :: :    :    :::
A A       CC TGCACTC      C        CA CAAGC  TT C      A    TGT

CACTGACCCTCTTCTCTGCACAGCTCCTAAGCCACTGCCTGCTGGTGACCCTGGCCGCCCACCTCCC
 : :  ::          :  :: :  ::  :::::::::  :::::::::  :::::::  ::: :  :    :::: :::
C CGGA        CAACT T CA  AGCTCCTGAGCCACTGCTTGCTGGTGA  CCTCGGCTAGCACCACCC

CCTGCGGT G CACGCTTCCCTGGACAAGTTCCTGGCTTCTGTGAGCACCGTGCTGACCTCCAAATA
::: :  : :     ::  : :::                :  :::     :  : ::  :::::::::::: ::
CCTCCCCTCGTGGTGCATGCCTCTCTTGACAAATTCCTTGCCTCTGTACTGTACTGACCTCCAAGTA

GCCTCGGTAGCCGTTCCTCCTGCCCGCTGGGCCTCCCAACGGGCCCTCCTCCCCTCCTTGCACCGGC
::::: : ::: :     :: :  :::       ::   ::      : :  :::: ::::: ::    :
GCCTC CT GCC AGCCT    TGCCTTCT GG CT      A GACCCTTCTTCCCTCCCTTGGATTCGT

TGAATAAAGTCTGAGTGGGCGGCAGCCTGTGTGTGCCTGGGTTCTCTCTGTCCCGGAATGTGCCAAC
:::::::::: :     :::::  : :      : : : :       : ::: ::       ::   : ::: : ::
TGAATAAAG C  CCTGGGC  C G    G G G G AT     CAT TCT TG       GG   TCTG C AC

ACCTGTCTCAGACCAAGGACCTCTCTGCAGCTGCATGGGGCTGGGGAGGGAGAACTGCAGGGAGTAT
 :  : : :  ::  :: :  :: :: : :::     :::: : :: ::  : ::   : ::: :
   GAAT A A  AA G CTTATTTG AG TTCAT    GCTGATG GGCAG A TGGA AG GTCT

GAGGTGGGCCTGCTCAAGAGAAGGTGCTGAACCATCCCCTGTCCTGAGAGGTGCCAGCCTGCAGGCA
 :  ::: : :    ::::  :: :::: ::    :         ::    :   : :: ::  :  : ::::
A  TGGAACAG    AAGA CAGATGCTTAA  A         TG AGAG CAGCTTTCA CC   CTGGCA
```

Figure 17. An alignment of human α-globin (HSAGL1) with a mouse pseudogene for α-globin (MMAGLX).

213

```
                                FQTV                       GSG      LDHILS            LA
HPMVP4  MNMSRQGI
TVGVP4  GNASSSDKSNSQS SGNE GVIINNFYSNQYQNSID LS            ASGGNAG  DAPQNNG  QLSSILGG AANAFATM APLLM
MENVP4  GNSTSSDKNNSSS EGNE GVIINNFYSNQYQNSID LS            ANAT GS  DPPKTYG  QFSNLLSG AVNAFSNM LPLLA
EMCVP4  GNSTSSDKNNSSS EGNE GVIINNFYSNQYQNSID LS            ANAA GS  DPPRLRS  IFESL SG AVNAFSNM LPLLA
F1KVP4  GAGQSSPATGSQNQSGNT GSIINNYMQQYQNSMDTQLGDNAISGGSNEGSTDTTSTHTTNTQNNDWFSKLASSAFSGLFGALLA APALN
H14VP4  MGAQVSTQKSGSHENQNILTNGSNQTFTVINYYKDAA              STSSAGQ  SLSMDPS  KFTEPVKDLMLKG     APALN
H89VP4  MGAQVSRQNVGTHSTQNSVSNGSSLNYFNINYFKDAA              SSGASRL  DFSQDPS  KFTDPVKDVLEKG     IPTLQ
PN2VP4  MGAQVSSQKVGAHENSNRAYGGSTINYTTINYYRDSA              SNAASKQ  DFAQDPS  KFTEPIKDVLIKT     APTLN
CB4VP4  MGAQVSTQKTGAHETGLNASGNSIIHYTNINYYKDAA              SNSANRQ  DFTQDPG  KFTEPVKDIMIKS     LPALN
BEVVP4  MGAQLSRNTAGSHTTGTYATGGSTINYNNINYYSHAA              SAAQNKQ  DFTQDPS  KFTQPIADVIKET     AVPLK
```

Figure 18. Multiple alignment of the picornavirus coat proteins VP4 by the method of Feng and Doolitle. HPM, Hepatitis A virus; TVG, Theiler's murine encephalomyelitis virus; MEN, Mengo virus; EMC, Encephalomyocarditis virus; F1k, Foot and mouth disease virus strain 0−1-k; H14, Human rhinovirus type 14; H89, Human rhinovirus type 89; PN2, Poliovirus; CB4, Coxsackievirus; BEV, Bovine enterovirus.

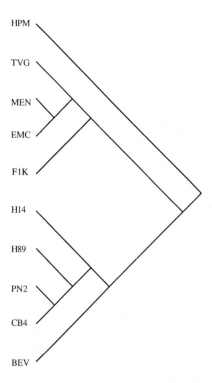

Figure 19. The tree used to build the multiple alignment of *Figure 18*. For abbreviations see the legend to *Figure 18*.

optimal alignments which for *n* sequences number $n(n-1)/2$. The most similar pairs are aligned and the next most similar sequence is then aligned with the alignment of the pairs (*Figure 18*). This process is continued until all the sequences have been included and implies a bifurcating tree relating the sequences (it could be an evolutionary tree, *Figure 19*). The method usually produces sensible results and is satisfactory in practice.

Multiple alignments give a picture of the possible variants in a family of proteins and can help in the search for further members of the family (33,34). A position-specific scoring table (or 'profile') is created for the group of aligned sequences and this profile is then used as a probe against a protein sequence database by an exhaustive alignment method.

7.6 **Predicting protein function**

The function of proteins is intimately related to their structure and sequence is usually less useful as a predictor of function. The structures of proteins which have been determined by X-ray crystallography are held in the Protein Data Bank of the Brookhaven National Laboratory. Useful insights may be obtained into the possible structure in the case that a protein is of similar sequence to a protein of known structure by using molecular graphics to model changes which might be expected because of differences in amino acid residues. This is a topic which is beyond the scope of this Chapter.

However, there are a number of methods which can give some useful information even in the absence of a structure.

7.6.1 *Hydropathy*

The behaviour of an amino acid residue in a protein is largely related to only two factors: primarily size of the residue and secondarily its charge. These parameters provide a means of ordering amino acids from hydrophilic (seeking contact with water molecules) to hydrophobic (avoiding water and seeking the interior of a protein or membrane) and this scale has been called hydropathy. The precise formulation of a hydropathy scale is somewhat arbitrary. Sweet and Eisenberg (35) showed that similar three-dimensional protein structures are correlated with similar hydropathic profiles. Hopp and Woods (36) and Kyte and Doolittle (37) plotted the hydropathy changes along protein sequences by the sliding window technique (Section 7.2) and such plots are useful for predicting B-cell antigenic determinants (hydrophilic surface residues), and signal peptides and membrane spanning regions (hydrophobic).

7.6.2 *Secondary structure prediction*

The structures of globular proteins are largely built out of α-helices, β-sheets, coils and turns (38) with the helices and sheets packing together in a variety of motifs. Many approaches to the prediction of protein structure from sequence have been investigated with only moderate success (5, *Figure 20*). One of the most satisfactory methods is that of Garnier, Osguthorpe and Robson (42) which quantifies the contribution which each residue makes to α-helix, β-sheet, coil and turn in a sliding window. The curves for each of the four states are plotted on a single axis and the highest at any position is the prediction of structure. The accuracy of the prediction is around 50%. Clearly, much better accuracy is obtained by the method of considering sequences similar to known structures (43).

7.6.3 *Sequence motifs*

Some protein binding or active sites are sufficiently well characterized by sequence alone that they can be located in the absence of structural information (44). In some cases their specification is complex and the motif must be found by a proportional algorithm. In others a relatively simple pattern with some ambiguities enables these motifs to be located extremely efficiently (*Table 6*).

8. PROSPECT

Molecular sequences are the genetic information which is interpreted by the actions of their own transcribed and translated products. Computers are already indispensible for the acquisition, storage and analysis of the sequence and structure of these genetic functions. As our technical ability to determine sequences and structures improves the rate of acquisition will accelerate. Fortunately, developments in computer technology, especially optical storage media and parallel processing, are keeping step with this acceleration in biological data acquisition. The specific suggestions and recommendations made in this Chapter can be expected to have only a short life.

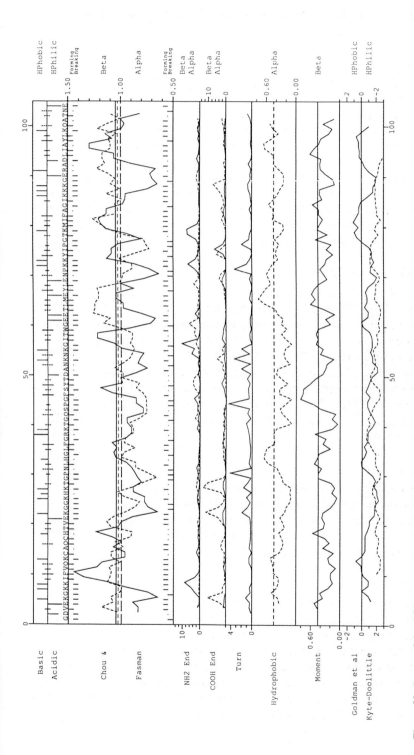

Figure 20. Output of the GCG program Pepplot for hippopotamus cytochrome *c*. At the top is a classification of the residues indicated by the length of the lines as basic, acidic, hydrophobic or hydrophilic with the sequence below. There follows a secondary structure prediction by the Chou and Fasman method (39) showing propensity for α-helix and β-sheet formation and indication of residues that are β-sheet forming or breaking and α-helix forming or breaking. Next are regions found at the N terminus of α and β structures, the C terminus of α and β structures, and regions found in turns. Plots relating to α helices are shown as dotted lines, β sheets as unbroken ones. Next is shown hydrophobic moment according to Eisenberg *et al.* (40) and finally hydropathy according to Engelman *et al.* (see 41) and Kyte and Doolittle (37). Plots relating to α-helices are shown as dotted lines, β-sheets as unbroken ones.

217

Table 6. Protein sequence motifs in the single letter code.

N-glycosylation	NX(ST)
Purine binding site	GXXXXGK(ST)
Nuclear protein transit sequence	KKKRKV
Factor X protease cleavage site	IEGR
Viral polyprotein cleavage site	(QY)G
Serine protease active site	GDSGG
Cysteine protease active site	GDCGG
Acid protease active site	FDTGS
Phosphotransferase active site	HXDhXXXXNhhh
Fibronectin cell adhesion sequence	RGDS
DNA binding site	CXXCXXXXHXXXXC
T-cell antigenic site	(DEKHRG)(hAT)(hAT)(hATDEKHRNQSPG)

X is any amino acid and residues in parentheses are alternatives at a single site. A hydrophobic residue (FLIMVWY) is indicated by h.

9. REFERENCES

1. Martin,W.J. and Davis,R.W. (1986) *BioTechnology*, **4**, 890.
2. Staden,R. (1987) In *Nucleic Acid and Protein Sequence Analysis: A Practical Approach*. Bishop,M.J. and Rawlings,C.J. (eds), IRL Press, Oxford, p. 173.
3. Gouy,M. (1987) In *Nucleic Acid and Protein Sequence Analysis: A Practical Approach*. Bishop,M.J. and Rawlings,C.J. (eds), IRL Press, Oxford, p. 259.
4. Stormo,G. (1987) In *Nucleic Acid and Protein Sequence Analysis: A Practical Approach*. Bishop,M.J. and Rawlings,C.J. (eds), IRL Press, Oxford, p. 231.
5. Taylor,W.R. (1987) In *Nucleic Acid and Protein Sequence Analysis: A Practical Approach*. Bishop,M.J. and Rawlings,C.J. (eds), IRL Press, Oxford, p. 285.
6. Bishop,M.J., Ginsburg,M., Rawlings,C.J. and Wakeford,R. (1987) In *Nucleic Acid and Protein Sequence Analysis: A Practical Approach*. Bishop,M.J. and Rawlings,C.J. (eds), IRL Press, Oxford, p. 83.
7. Kohara,Y., Akiyama,K. and Isono,K. (1987) *Cell*, **50**, 495.
8. Collins,J.F. and Coulson,A.F.W. (1987) In *Nucleic Acid and Protein Sequence Analysis: A Practical Approach*. Bishop,M.J. and Rawlings,C.J. (eds), IRL Press, Oxford, p. 323.
9. Roode,D., Liebschutz,R., Maulik,S., Friedemann,T., Benton,D. and Kristofferson,D. (1988) *Nucleic Acids Res.*, **16**, 1857.
10. Bicknell,E.J., Rada,R., Davidson,S. and Stander,R. (1988) *Nucleic Acids Res.*, **16**, 1667.
11. Elder,J.K. and Southern,E.M. (1987) In *Nucleic Acid and Protein Sequence Analysis: A Practical Approach*. Bishop,M.J. and Rawlings,C.J. (eds), IRL Press, Oxford, p. 219.
12. Staden,R. (1988) *CABIOS*, **4**, 53.
13. Doolittle,R.F. (1986) *Of URFs and ORFs. A Primer on How to Analyse Derived Amino Acid Sequences*. University Science Books, Mill Valley, California.
14. Coulondre,C., Miller,J.H., Farabaugh,P.J. and Gilbert,W. (1978) *Nature*, **274**, 775.
15. Staden,R. (1985) In *Genetic Engineering: Principles and Methods*. Setlow,J.K. and Hollaender,A. (eds), Vol. 7, p. 67.
16. Grantham,R., Gautier,C., Gouy,M., Mercier,R. and Pave,A. (1980) *Nucleic Acids Res.*, **8**, r49.
17. Ikemura,T. (1981) *J. Mol. Biol.*, **146**, 1.
18. Konigsberg,W. and Godson,G.N. (1983) *Proc. Natl. Acad. Sci. USA*, **80**, 687.
19. Anderson,S., Bankier,A.T., Barrell,B.G., de Bruijn,M.H.L., Coulson,A.R., Eperon,I.C., Sanger,F. and Young,I.G. (1982) *J. Mol. Biol.*, **156**, 683.
20. Alff-Steinberger,C. (1987) *J. Theor. Biol.*, **124**, 89.
21. Martinez,H.M. (1983) *Nucleic Acids Res.*, **11**, 4629.
22. Karlin,S., Ghandour,G., Ost,F., Tavare,S. and Korn,L.J. (1983) *Proc. Natl. Acad. Sci. USA*, **80**, 5660.
23. Moore,G.P. and Moore,A.R. (1982) *J. Theor. Biol.*, **98**, 165.
24. Houck,C.M., Reinhart,F.P. and Schmid,C.W. (1979) *J. Mol. Biol.*, **132**, 289.
25. Bird,A.P. (1987) *Trends Genet.*, **3**, 342.
26. Roberts,R.J. (1987) *Nucleic Acids Res.*, **15**, r189.
27. Wilbur,W.J. and Lipman,D.J. (1983) *Proc. Natl. Acad. Sci. USA*, **80**, 726.

28. Lipman,D.J. and Pearson,W.R. (1985) *Science*, **227**, 1435.
29. Pearson,W.P. and Lipman,D.J. (1988) *Proc. Natl. Acad. Sci. USA*, **85**, 2444.
30. Needleman,S.B. and Wunsch,C.D. (1970) *J. Mol. Biol.*, **48**, 443.
31. Barton,G.J. and Sternberg,M.J.E. (1987) *J. Mol. Biol.*, **198**, 327.
32. Feng,D.F. and Doolittle,R.F. (1987) *J. Mol. Evol.*, **25**, 351.
33. Patthy,L. (1987) *J. Mol. Biol.*, **198**, 567.
34. Gribskov,M., McLachlan,A.D. and Eisenberg,D. (1987) *Proc. Natl. Acad. Sci. USA*, **84**, 4355.
35. Sweet,R.M. and Eisenberg,D. (1983) *J. Mol. Biol.*, **171**, 479.
36. Hopp,T.P. and Woods,K.R. (1981) *Proc. Natl. Acad. Sci. USA*, **78**, 3824.
37. Kyte,J. and Doolittle,R.F. (1982) *J. Mol. Biol.*, **157**, 105.
38. Chothia,C. (1984) *Annu. Rev. Biochem.*, **53**, 537.
39. Chou,P.Y. and Fasman,G.D. (1978) *Adv. Enzymol.*, **47**, 45.
40. Eisenberg,D., Weiss,R.M. and Terwilliger,T. (1984) *Proc. Natl. Acad. Sci. USA*, **81**, 140.
41. Engelman,D.M., Steitz,T.A. and Goldman,A. (1986) *Annu. Rev. Biophys. Chem.*, **15**, 321.
42. Garnier,J., Osguthorpe,J.D. and Robson,B. (1978) *J. Mol. Biol.*, **120**, 97.
43. Chothia,C. and Lesk,A.M. (1986) *Proc. R. Soc. Lond. A*, **317**, 345.
44. Hodgman,T.C. (1986) *Comp. Appl. Biosci.*, **2**, 181.

CHAPTER 8

Automated DNA sequencing

CHERYL HEINER and TIM HUNKAPILLER

1. INTRODUCTION

DNA sequencing has undergone great advances since the initial developments of Maxam and Gilbert (1) and Sanger *et al.* (2) in the 1970s. The Maxam−Gilbert or chemical method first employs specific chemicals to modify DNA at one or two bases. The DNA is then cleaved at the modified site to generate fragments shorter by one base than the targeted base. In the Sanger or dideoxy method, a specific primer is annealed and enzymatically extended to synthesize a complement of the unknown sequence. The reaction mixtures contain the four deoxynucleoside triphosphates (dNTPs) plus one dideoxynucleoside triphosphate (ddNTP). Incorporation of the dideoxy analogue inhibits further extension of the synthesis. By carefully titrating the reagents and substrate, both methods ensure that every possible base position is the site of either cleavage or strand elongation termination. Therefore, a complete nested set of fragments is generated. Both methods use high resolution polyacrylamide gel electrophoresis to separate these fragments. Until recently, radioactive markers were the only method of labelling these fragments, with autoradiography the method of detection. Sets of reactions are loaded onto adjacent gel lanes, one reaction per lane, and the sequence is determined by reading bands going up the autoradiograph. In this manner 300 bases or more can be determined from a given fragment.

The M13 cloning−dideoxy sequencing system has gained widespread popularity in recent years, in large part due to the development of modified M13 bacteriophage as cloning vectors by Messing and coworkers (3) (see Chapter 1). The M13mp vectors facilitate insertion of restriction fragments, colour selection of recombinant plaques, and production of relatively high quantities of single-stranded DNA. In addition, a universal primer complementary to the phage DNA at a site close to the phage/insert sequence boundary can be used for sequencing purposes.

Along with the generation of new M13 vectors, many other techniques have recently been developed which expand the utility, decrease time and/or improve results of dideoxy sequencing. Sequencing of double-stranded DNA (see Chapter 4), especially plasmids, eliminates the need to subclone into M13 (4), and may give results comparable to those obtained from single-stranded templates. Large inserts can be sequenced by gene walking (5), which involves making one or more primers which hybridize to sequence within the insert, once some of the sequence has been determined (see Chapter 2, Section 1.2.2). Alternatively, large inserts may be partially digested to generate a nested set of deletion clones (6) (see Chapter 1, Section 3.3.7). Both of these 'directional' sequencing methods can greatly reduce the total number of sequencing reactions required for a given project.

The enzyme used originally for extension of primers was the Klenow fragment of

DNA polymerase I. Reverse transcriptase proved useful for sequencing certain DNA sequences and for sequencing RNA by the dideoxy method (see Chapter 6, Section 5). Then a modified phage T7 polymerase became popular for sequencing (7) (see Chapter 2, Section 3.3). It has the advantage of generating a more even distribution of fragments than Klenow or reverse transcriptase, making data interpretation much easier. Recently, Taq polymerase has become available; it allows polymerization to occur at high temperatures ($60° - 80°C$), and generates a nearly uniform, high concentration of products. GC-rich sequences often have anomalous migration on acrylamide gels, making sequences in such regions difficult to determine. This problem has been lessened with the use of G analogues in sequencing reactions, such as inosine and 7-deaza-guanosine nucleotides (see Chapter 2, Section 3.2).

Accompanying these improvements in cloning and sequencing techniques, there have been refinements in the second half of the process, which includes separation of the fragments in a high resolution denaturing polyacrylamide gel, detection of the resolved fragments, and interpretation of the results. Longer (see Chapter 5, Section 9) and thinner gels provide better resolution of larger fragments, although there are some technical challenges in making them. Likewise, wedge or gradient gels (see Chapter 2, Section 5.7) allow for more even spacing of bands and thus more bands per gel, but add to pre-sequencing preparation efforts. The obvious first choice as a label was ^{32}P, since ^{32}P-labelled dNTPs were readily available and are incorporated easily by polymerases. Now ^{35}S is available as an alternative label and can provide significantly improved resolution. However, use of ^{35}S requires the additional step of gel drying and significantly longer exposure times.

More recently, fluorescent dyes have become an alternative to radiolabels and have been used as part of automated systems using laser excitation and direct optical detection. In these systems four dyes with different emission frequencies can be used as the labels, one for each different reaction. The products of all four reactions are then combined and electrophoresed on one lane of a gel. The fluorescently-labelled bands are detected in real time as they migrate past a detector near the bottom of the gel. The signal spectrum is then analysed to determine which dye (i.e. which base) the bands represent. Currently, dye-labelled sequencing is available only for dideoxy reactions. The dyes may be attached to the 5′ end of the primer (8), or to the terminating nucleotide analogue (9). The protocols outlined here pertain to the labelled primer approach.

2. PROTOCOL

Determining a nucleic acid sequence is a complex process with many steps, which currently involves days of work. With manual sequencing, the following steps are generally required.

(i)   Purification of unknown DNA from original organism or from library.
(ii)   Fractionation of DNA into suitable size fragments ($300 - 5000$ bp).
(iii)   Digestions, ligations, and transformation using an M13 vector and host.
(iv)   Growth and purification of M13 recombinant DNA.
(v)   Dideoxy sequencing reactions.
(vi)   Polyacrylamide gel electrophoresis, gel disassembly and exposure.
(vii)   Reading of autoradiograph, and entering of data into a computer.

(viii)   Comparing data from different gels and clones for overlapping sequences or to determine a consensus sequence.

The first tools to help automate this process were computer programs designed to compare and meld different sequences, greatly assisting in step (viii) above. There are many programs currently available which perform this function (see Chapter 7). More recently tools to facilitate the data entry of step (vii) have become more common, incorporating digitizing pads for direct, manual entry of gel data. Scanning systems and programs have also been developed to read and interpret autoradiographs automatically. However, the latter have not been able to perform as well as manual reading to date. There has also been considerable effort to develop pre-cast, disposable acrylamide gels in order to eliminate the tedious and technique-oriented job of pouring gels. Such gels are not commonly available yet. The most revolutionary efforts at automating the process of sequencing to date relate to the automated system of detection and analysis using fluorescently-labelled primers discussed above, and will be the focus of the remainder of this Chapter. The process, developed by research at the California Institute of Technology (10) and Applied Biosystems, Inc. (11), provides a means to improve greatly the data throughput of any sequencing project, as well as reduce costs, safety risks and error due to manual data entry. The following protocols are based on and support the ABI 370A automated DNA sequencer. The protocols for fluorescent sequencing are representative for M13-based sequencing, although plasmid sequencing may also be done with this system.

2.1 M13 cloning and purification

Cloning strategies are set out in Chapter 1. Once an approach has been decided upon, the DNA must be inserted into M13 RF or other suitable vector DNA, and the recombinant DNA transformed into an appropriate strain of *Escherichia coli*. Recombinant plaques are then picked and grown up for a limited time. Single-stranded DNA is extruded from the cells as phage particles into the medium, without lysing the bacteria. It is then a relatively simple matter to purify the M13 phage particles away from intact cells. The protein coat is stripped from the phage by phenol extraction, and the DNA is concentrated by ethanol precipitation. Protocols for constructing recombinant phage are given in Chapter 1. Single-stranded DNA is prepared using a modification outlined below of the standard protocol in Chapter 2.

2.1.1 *Reagents*

2 × TY medium:

Bactotryptone	16.0 g	
Yeast	5.0 g	
NaCl	5.0 g	per litre of distilled H_2O
pH 7.2−7.4		

PEG solution: 20% polyethylene glycol 8000, 2.5M NaCl
TE buffer: 10 mM Tris−HCl, pH 8.0, 1 mM Na_2EDTA
All media, solutions, buffers, test tubes and pipettes should be sterile.

2.1.2 *Protocol*

(i)   Inoculate 10 ml of 2 × TY medium with a single colony of *E. coli* grown on glucose minimal agar. Shake overnight at 37°C.

(ii)   Dilute a portion of the overnight culture 1:100 in 2 × TY medium; aliquot 5 ml into 10 ml, or larger, tubes or flasks.

(iii)   Pick recombinant (colourless) plaques with a sterile toothpick or pipette; choose only well-separated plaques. Inoculate each culture with a single plaque.

(iv)   Grow for 6−7 h with vigorous shaking at 37°C. *Note*: Longer growth may result in bacterial lysis and contamination of M13 DNA with cellular DNA, RNA, and debris.

(v)   Pellet cells by centrifugation at 2500 r.p.m. for 10 min, transfer supernatants to fresh tubes. (Pellets may be used for preparation of double-stranded RF DNA.) Cultures may be stored at this stage for several days at 4°C. Repeat low-speed spin to remove any residual cells, even if cultures have not been stored.

(vi)   Precipitate phage particles by adding 1 ml of PEG solution; incubate at 4°C for 30 min. Collect phage by spinning at 10 000 r.p.m. for 30 min. Discard supernatants; remove the last traces of PEG by allowing the tubes to drain upside down for several minutes. Resuspend the pellets in 400 μl of TE buffer; transfer to 1.5-ml microcentrifuge tubes.

(vii)   Extract with an equal volume of phenol which has been equilibrated with TE buffer. Vortex to mix, then spin for a few minutes in a microcentrifuge to separate the phases. Transfer the upper aqueous phase to a fresh tube, being careful not to disturb the interface.

(viii)   Repeat extraction with 200 μl of equilibrated phenol plus 200 μl of chloroform:isoamyl alcohol, 24:1.

(ix)   Extract with an equal volume of ether to remove traces of phenol. Remove the upper ether layer, and repeat the ether extraction. Spin under vacuum for a few minutes to remove residual ether.

(x)   Ethanol-precipitate by adding 0.1 vol of 3 M sodium acetate, pH 5.2, plus 3 vol of 95% ethanol. Precipitate for at least 20 min at −70°C, or overnight at −20°C. Spin for 20 min at 4°C. Discard supernatant, and rinse pellet with 1 ml of 70% ethanol. Spin for an additional 5 min, and discard the supernatant. Drain upside down or draw off residual ethanol with a pipette. Remove the last traces of ethanol by drying in a vacuum centrifuge. Resuspend the pellet in 20−50 μl of TE buffer.

(xi)   To check size and nucleic acid purity, run 1−2 μl on a 0.8% agarose gel with M13 (no insert) as a standard, stain with ethidium bromide, and visualize under UV light. The DNA should run as a single sharp band; look for a diffuse band of RNA near the bottom of the gel, and bacterial DNA near the wells. If either of these is present, results are likely to be poor.

(xii)   Measure the absorbance at 260 and 280 nm. The 260:280 ratio should be 1.65 or greater. Calculate DNA concentration, an important step in obtaining good results with fluorescent labels in sequencing. Since each band is only being detected for a few seconds in real time, there must be enough fluorescence present to yield the required signal. There is no chance to go back and do a 'darker exposure'. Note: 1 O.D. unit = 33 μg SS DNA (approximately). 1 O.D. unit is the quantity of DNA which has an absorbance of 1 at 260 nm measured in a cuvette of 1 cm path length, when the sample volume is 1 ml.

2.2 Dideoxy sequencing reactions

2.2.1 *Background*

Sequencing reactions with dye-labelled primers involve three steps:

(i)   Annealing of primer to template, each reaction in a separate tube.
(ii)   Extension of primer with a polymerase and the four dNTPs plus one ddNTP.
(iii)   Combining the four reactions for one template into one tube and concentrating by ethanol precipitation.

There are several different protocols involving different enzymes, templates and reaction conditions currently in use. Two of the most commonly used protocols are given below. Each protocol yields sufficient quantity for two loadings. Reactions may be scaled up or down according to need.

2.2.2 *Reagents*

(i)   *10 × Sequencing buffer (seq. buffer)*. 100 mM Tris−HCl, pH 7.5, 100 mM MgCl$_2$, 500 mM NaCl.
(ii)   *Dye-labelled primers*. Dissolve freeze-dried primers in sterile TE buffer (see Section 2.1.1) to a concentration of 0.4 pmol μl^{-1}. Vortex thoroughly. Store dye primers at −20°C in the dark. Note: Dye-labelled DNA is mildly light sensitive. Avoid excessive exposure to light.
(iii)   *Template*. Purify and quantitate template as described in Section 2.1.2.
(iv)   *Klenow dNTP mixes*. A separate mix is made for each reaction

      A mix: 50 μM dATP, 500 μM dCTP, 500 μM dGTP, 500 μM dTTP
      C mix: 500 μM dATP, 50 μM dCTP, 500 μM dGTP, 500 μM dTTP
      G mix: 500 μM dATP, 500 μM dCTP, 50 μM dGTP, 500 μM dTTP
      T mix: 500 μM dATP, 500 μM dCTP, 500 μM dGTP, 50 μM dTTP

      All mixes are made in sterile distilled water.

(v)   *Klenow ddNTP mixes*. Again a separate dilution is made of each ddNTP in sterile distilled water as follows: 0.6 mM ddATP, 0.2 mM ddCTP, 0.3 mM ddGTP, 1.0 mM ddTTP.
(vi)   *Klenow d/ddNTP mixes*. Combine dNTP mixes with the appropriate ddNTP in a 1:1 ratio for use in sequencing reactions.
(vii)   *Sequenase dNTP mix*. The same dNTP mix is used for all four reactions, containing 2 mM of each dNTP in distilled water.
(viii) *Sequenase ddNTP mixes*. Make a separate dilution for each ddNTP, at a concentration of 50 μM in distilled water.
(ix)   *Sequenase d/ddNTP mixes*. Combine dNTP mix with the appropriate ddNTP in a 1:1 ratio. *Note*: Band compression in GC-rich sequences is minimized by using dc7GTP in place of dGTP. Use dc7GTP in all four reactions at 1.5 × the concentration of dGTP (i.e. use 750 or 75 μM dc7GTP for Klenow mixes, or 3 mM dc7GTP for Sequenase).

(x)   *DTT/Sequenase mix.* (Combine immediately before use) 12−13.5 units of Sequenase added to 4.5 µl of 0.1 M dithiothreitol plus cold TE buffer, in a final volume of 9 µl. This is sufficient for one set of reactions as outlined below. Scale up or down according to need. Use extra enzyme for problematic sequences.

(xi) *Taq 10 × sequencing buffer.* Same as (i), except pH 8.5.

(xii) *Taq dNTP mixes.* Same as (iv), except the limiting dNTP (dATP for A mix, etc.) is at 125 µM.

(xiii) *Taq ddNTP mixes.* A separate dilution is made of each ddNTP in sterile distilled water as follows: 3.0 mM ddATP, 1.5 mM ddCTP, 0.25 mM ddGTP, 2.5 mM ddTTP.

(xiv) *Taq d/ddNTP mixes.* Combine dNTP mixes with the appropriate ddNTP in a 1:1 ratio for use in sequencing reactions.

2.2.3 Klenow reaction protocol

Note: Twice as much template is required for the G and T reactions as for the A and C reactions due to the different relative fluorescence of the dyes.

(i) *Annealing*

Reagent	A rtn	C rtn	G rtn	T rtn
Distilled water	q.s.	q.s.	q.s.	q.s.
10 × seq. buffer (µl)	1.5	1.5	3.0	3.0
Dye primer (µl)	2.0	2.0	4.0	4.0
(0.4 pmol µl^{-1})	JOE	FAM	TAMRA	ROX
Template DNA (pmol)	0.4	0.4	0.8	0.8
Total volume (µl)	12.0	12.0	24.0	24.0

Combine the above reagents in the order listed in labelled Eppendorf tubes and centrifuge briefly to bring the contents to the bottom of the tube. Incubate at 55°C for 5 min then cool slowly over 30 min to room temperature. (A heater with a removable block works well; place the block in the refrigerator or on the bench after the 55°C incubation.)

(ii) *Extension reactions*

Add to the cooled tubes of annealed DNA:

Reagent	A rtn	C rtn	G rtn	T rtn
d/ddNTP mix (µl)	2.0	2.0	4.0	4.0
Klenow polymerase (units)	4.0	4.0	8.0	8.0
Total volume (µl approx)	15.0	15.0	30.0	30.0

Centrifuge briefly to bring the contents to the bottom of the tube, if necessary. Incubate under one of the following conditions:

(a)   16°C for 30 min (recommended if there is possibly a second site on the template with considerable complementarity to the primer.

(b)   Room temperature (20−25°C) for 15 min.

(c)   37°C for 10 min.

(d)   42°C for 10 min (recommended only for GC-rich sequences or sequences with much self-complementarity).

At the end of the incubation, heat the reactions to 65°C for 10 min to inactivate the enzyme. Reactions may be stored at this stage for several weeks at −20°C in the dark.

(iii) *Ethanol precipitation*

Combine half of each sequencing reaction for one loading as follows: 7.5 μl of A rtn, 7.5 μl of C rtn, 15 μl of G rtn, 15 μl of T rtn to a total of 45 μl. Add 4.5 μl of 3 M sodium acetate, pH 4.5, and 150 μl of cold 95% ethanol. Incubate on ice for 10−20 min. Spin for 20 min at 4°C. Pour off the supernatant, and rinse the pellet with 500 μl of cold 70% ethanol. Spin again for 2 min at 4°C, being careful to maintain the orientation of the pellet if using a fixed angle rotor. Pour off the supernatant; remove extra ethanol by blotting upside down on a tissue or with a pipette. Dry briefly in a vacuum centrifuge. Reactions may be stored at this stage for several months at −20°C in the dark.

2.2.4 *Sequenase reactions*

(i) *Annealing*

Reagent	A rtn	C rtn	G rtn	T rtn
Distilled water	Sufficient to make up final volume			
5 × seq. buffer (μl)	1.0	1.0	3.0	3.0
Template DNA (pmol)	0.2	0.2	0.6	0.6
Dye primer (μl)	1.0	1.0	3.0	3.0
(0.4 pmol μl^{-1})	JOE	FAM	TAMRA	ROX
Total volume (μl)	5.0	5.0	15.0	15.0

Add reagents to Eppendorf tubes in the order listed, mix, and spin briefly to bring the contents to the bottom of the tube. Incubate at 55°C for 5 min. Cool to room temperature for 30−60 min.

(ii) *Extension reaction*

Reagent	A rtn	C rtn	G rtn	T rtn
d/ddNTP mix (μl)	1.5	1.5	4.5	4.5
DTT/Sequenase mix (μl)	1.0	1.0	3.0	3.0
Final volume (μl)	7.5	7.5	22.5	22.5

Add reagents to annealed DNA in the order listed, pipetting up and down after each addition. DNA and NTPs should be at room temperature. After adding the enzyme mix, immediately place the tube in a 37°C block or bath and incubate for 5−7 min. Inactivate the enzyme by heating to 65°C for 10 min. Reactions are stable at this stage for several weeks if stored at −20°C in the dark.

(iii) *Ethanol precipitation*

Combine half of each reaction for one loading as follows: A and C reactions, 3.5 μl; G and T reactions, 10.5 μl to give a final volume of 28 μl. Add 2.8 μl of 3 M NaOAc plus 90 μl of cold 95% ehtanol, and precipitate on wet ice for 10−20 mins. Continue as in step (iii) of Section 2.2.3.

2.2.5 *Taq reactions*

Reactions using Taq polymerase are currently being optimized. Very good results have been obtained following a protocol similar to that for Klenow, with the following changes.

(i) *Annealing*

(a) Use Taq 10 × sequencing buffer.

(b) Half as much template, or less, may be used.

(ii) *Extension*

(a) Use Taq d/ddNTP mixes and Taq polymerase.

(b) Incubate at 70°C for 5 to 10 minutes.

(c) Omit the heat inactivation step. (It does not work for Taq polymerase.)

(iii) *Ethanol precipitation*

Add reactions directly to ethanol, as the enzyme has not been inactivated.

2.3 Polyacrylamide gel preparation

2.3.1 *Background*

The dye-labelled fragments produced in the sequencing rections are separated on high-resolution polyacrylamide gels, which are capable of resolving differences in length of one base. Electrophoresis is carried out under denaturing conditions to minimize sequence-specific secondary structures which cause anomalous migrations, as well as to prevent the extended primers from reannealing to the template DNA. High resolution is achieved by use of thin gels (0.4 mm or less) and requires the use of high quality reagents and materials. The gel and surrounding glass plates must be free of dust, dirt and fluorescent contaminants which can confuse data interpretation and lead to poor results. The protocol outlined below generates a 6% polyacrylamide gel, which gives at least 400 bases of readable sequence from the primer. Different composition gels may be desirable for better resolution closer to the primer (8%), or further downstream (4%).

2.3.2 *Reagents*

40% acrylamide stock solution (19:1 acrylamide:bis-acrylamide):

Acrylamide 380 g

N,N'-methylenebisacrylamide (bis-acrylamide) 20 g

Dissolve in distilled water, bring to 1 l final volume. Store at 4°C. Deionize for long-term storage.

CAUTION: Acrylamide and bisacrylamide are neurotoxins. Avoid inhalation of powder by weighing in a fume hood and avoid skin contact by wearing gloves whenever working with acrylamide powder or solutions. Dispose of waste safely.

10 × TBE buffer:

Trizma base   107.8 g

Boric acid   55.0 g

Na_2EDTA   9.3 g

Dissolve in distilled water, bring to 1 l final volume. Store at 4°C. Discard if a white precipitate appears.

2.3.3 *Preparation of a 6% denaturing polyacrylamide gel*

(i)   Weigh out 50 g of urea.

(ii)   Add 15 ml of 40% acrylamide stock solution, 10 ml of 10 × TBE, and enough distilled water to make approximately 100 ml. Stir (low heat will hasten dissolution) until urea is dissolved, then bring to 100 ml. Filter (to remove dust particles) and degas (for even, reproducible polymerization). While the urea is dissolving, prepare glass plates by thoroughly washing them with laboratory detergent, rinsing well with tap water, then deionized water, and finally 95% ethanol. Allow to air dry. Silanizing is not necessary.

(iii)   Assemble glass plates by laying one plate down on a clean flat surface and placing one spacer (0.4 mm thick) along each long edge. Place the second plate on top of the spacers, and line up the edges of both plates and the spacers. Tape all except the notched edge with wide sealing tape. Tape bottom edges twice. Clean a 0.4 mm thick comb by rinsing with distilled water.

(iv)   When the plates and gel solution are prepared, the gel may be poured. Polymerization is initiated by the addition of 500 μl of 10% (w/v) ammonium persulphate, stored at 4°C, less than one week old, and 50−75 μl of TEMED. Once the TEMED has been added, polymerization will occur within 5−10 min, so work quickly. A warm gel solution will speed up polymerization; cool to room temperature if this is the case. After adding ammonium persulphate and TEMED, swirl gently for a few seconds, then pour the solution between the glass plates, holding the plates at a slight angle from the vertical. A funnel or syringe may aid in pouring. Try to avoid getting bubbles in the solution between the plates; if they occur, hold the plates upright and wait for the bubbles to rise to the surface. If plates are not perfectly clean, bubbles may stick, although it may be possible to dislodge them by tapping the plates gently. As solution nears the top of the plates, lay the gel down horizontally on a block or with the top over the edge of the bench. Place the comb between the plates, making certain there are no air bubbles between the comb and the gel solution. Place several loose bulldog clamps over the teeth of the comb to ensure that no acrylamide accumulates between the comb and the glass. Allow the gel to polymerize for at least 90 min before use. If the solution leaks out from the bottom of the plates while pouring the gel, lay them nearly horizontal and continue working. If this is a common problem first pour a plug of quickly-polymerizing solution to seal the bottom of the gel as follows:

Add 500 μl of ammonium persulphate to the entire gel solution, then remove 10 ml into a small beaker and add 30−40 μl of TEMED. Quickly swirl the solution and pour about 5 ml between the plates, holding them nearly vertical. This solution should polymerize very quickly; watch the remaining 5 ml in the beaker to tell when polymerization has occurred. Then add 50−75 μl of TEMED to the remaining solution and pour as above.

Gels poured in the preceding manner are usable for a couple of days; store at room temperature and seal the area around the comb to prevent drying out by pouring the remaining gel solution over this area and covering with plastic wrap.

2.4 Gel set-up and loading

After the gel has polymerized for at least 90 min, it is ready to set up for running. The procedure is outlined below.

(i)   Remove tape and clamps from plates. Lay gel down on a flat surface, notched plate up. Clean area around comb if necessary for viewing wells. Carefully remove the comb, using a single-edge razor blade to lift it slightly from the lower plate to allow air underneath it. Place the razor blade between the notched edge of the glass and the stepped comb surface, and gently ease the comb out of the gel.

(ii)   After the comb has been removed, rinse the outside of the plates with warm tap water, then deionized water, and finally 95% ethanol. Flush out residual acrylamide from around the wells with deionized water. Lean plates against a wall to dry. Check the wells to see if any have bubbles underneath, residual acrylamide, or a bottom which is not flat. Any wells with these problems should be marked 'X' with a marker pen and not used. The other wells may be more easily visualized by outlining their edges.

(iii) Place the gel in the automated DNA sequencer and check plates for dirt, according to instrument menus. Reclean plates and check again if necessary. Prepare 1.8 l of 1 × TBE buffer from the 10 × stock (see Section 2.3.2). Position buffer chambers, tighten upper buffer chamber, and fill with buffer. Pre-run for 30−60 min at 30 W (power limiting), with the heater set to 40°C.

(iv)   While the gel is pre-running, prepare samples for loading. Add 5 μl of deionized formamide plus 1 μl of 50 mM Na_2EDTA (pH 8.0) to each precipitated reaction set; pipette up and down a couple of times to wash the sides of the tube. Spin briefly to bring the contents to the bottom of the tube. After the gel has pre-run, heat samples to 90°C for 2 min and chill briefly on ice. Meanwhile, flush the well with a syringe to remove urea, which continually leaches out of the gel and prevents the sample from layering evenly in the well. Then load the entire contents of one tube quickly into a well. A flattened gel loading tip (Section 4) or drawn-out capillaries will allow direct layering of the solution on the gel and generally give better resolution. Load only a few samples at a time, run them into the gel for 3−4 min, then load another set. Turn on electrophoresis immediately after loading each set of samples, and run at 30 W. Electrophoresis settings may be altered for different gel compositions, different buffer systems, or GC-rich samples which may require higher temperatures to denature secondary structures. (*Note*: Higher temperatures may result in decreased resolution).

(v)   Enter information about each sample, and begin run according to the ABI 370A software protocols.

3. DATA ANALYSIS

The data generated by this type of automated detection system differs from the standard autoradiograph in several ways. First, data is collected at one point in space over several hours of time. An autoradiograph looks at the entire space of the gel at one point in time. Thus, the first peaks of data correspond to the fastest moving fragments, which appear at the bottom of an autoradiograph. Second, this system combines four reactions in one lane; each reaction is labelled with a different fluorescent dye and is represented by a different colour. In radioactive systems, by contrast, each reaction is run on a separate lane, and the sequence read by comparing neighbouring lanes.

The raw data generated during the run represents the amount of fluorescence passing through each of four colour filters, amplified by a photomultiplier tube, for any lane or point on the gel from the beginning to the end of the run. The major steps in data analysis are outlined below.

(i) *Baseline subtraction.* The average baseline in each colour is subtracted from the raw data in that colour to bring all baselines to the bottom of the frame.

(ii) *Multicomponent analysis.* This system uses four filters for collection, each filter passing light around the maximum emission wavelength of one of the dyes. Ideally, each filter would pass light for only one dye. However, due to the fact that the emission spectra are roughly bell-shaped curves, in some cases there is spillover of light from one dye through the optimum filter for another dye. Thus, a pure band labelled with one dye may give rise to peaks in several colours, representing the emission spectrum of that dye through the four filters. Since the amount of spillover for each dye is known, it is relatively simple (for a computer) to determine the concentration of each dye at every point in time. In algebraic terms, it is a problem of solving four simultaneous equations for four unknowns. This type of system can determine whether one, two, three or all four dyes are present at any point, that is, it can separate overlapping peaks in two or more colours. This is helpful in resolving areas of band compression and in interpreting data when an underlying 'ghost sequence' is present.

(iii) *Mobility shifts.* The addition of a fluorescent dye to a fragment of DNA retards the mobility of that fragment in the gel. Since the four dyes used vary slightly in molecular weight and charge, it is not surprising that there are minor differences in relative mobilities between the dyes. The mobility correction is not necessarily essential for manual interpretation of the data, but the even spacing it provides assists in automatic base calling, in determining areas of band compression and interpretation of data at high base numbers. The mobility shifts apparently depend not only on the dye, but also on the sequence of bases near the dye, indicating possible dye−base interaction in the gel. For a given primer, the shifts will be constant, but may be slightly different with another primer sequence. These shifts also vary with different concentrations of acrylamide, although they remain relatively constant in the 4−8% range. All these variations are minor and can be adjusted with the appropriate software, but knowledge of their presence provides insight into the complexities of this type of system.

(iv) *Resolution enhancement.* This algorithm separates shoulders of peaks, and is especially helpful at high base numbers where resolution is not as good.

(v) *Scaling.* Each colour has its own scaling factor, depending on the relative intensity of that dye in the system.

(vi) *Base calling.* Peaks are located by size, shape, and spacing, and the colour or base is determined. Problem areas where peaks in more than one colour occur simultaneously or spacing is abnormal are flagged with a question mark. The obvious goal is for the computer to call bases as well as or better than an experienced person can do by hand. This is a difficult task, considering the variations in gels and run speed, sequence- and dye-specific spacing abnormalities, and variations in peak amplitudes from run to run, from colour to colour, from beginning to end of run, and even from differential rates of enzyme incorporation of a ddNTP at certain sequences. Improvements in this area can be expected over the next few years. However, well run reactions currently provide 400−500 bases of reliable data.

4. MATERIALS AND SUPPLIERS

(i)   Klenow fragment, DNA Polymerase I: Promega Biotech, International Biotechnologies, Inc., and others.

(ii)   Sequenase: Sequenase is the trademark of United States Biochemical Corporation for a chemically-modified T7 DNA polymerase. Pharmacia also market a T7 DNA polymerase.

(iii)   d/ddNTPs: Pharmacia, Boehringer-Mannheim Biochemicals.

(iv)   7-deaza-dGTP: Boehringer-Mannheim Biochemicals, Pharmacia.

(v)   Dye-labelled primers: Applied Biosystems, Inc.

(vi)   Acrylamide gel reagents: Ultra-pure reagents, especially acrylamide, bis-acrylamide, and urea, are essential. Many suppliers sell reagents of sufficient quality; check specifications.

(vii)   Gel loading tips: C.B.S. Scientific, Del Mar, CA.

(viii) Taq polymerase: Promega Biotech, Perkin Elmer, Cetus, and others.

5. TROUBLESHOOTING

The following list is by no means comprehensive, but represents some of the most common problems encountered in automated sequencing. Most of the problems are not specific to automated systems (see Chapter 3), although their manifestation may be different.

5.1 Extra or overlapping peaks

There are many possible causes of this problem, which leads to uninterpretable data or ?'s in automated base calls. Examine extra peaks to identify:

(i)   Second (and maybe third) sequence underlying the 'primary' sequence. This may be due to dirty template, that is, template contaminated with cellular nucleic acids or with M13 from another plaque. The contaminating nucleic acid should generally be visible on an ethidium bromide-stained agarose gel of the sample. Repurification is necessary if this is the problem.

    Another possible cause of an underlying sequence is the presence of a second priming site on the template. A computer search should be done on all new primers made to check for complementarity to the vector used. However, even if this turns up negative, there may exist second priming sites in the insert itself. One possible solution is to use Taq polymerase, as reactions are done under much more stringent hybridization conditions (70°C), greatly reducing second-site annealing. There are also two modifications of the Klenow or Sequenase reaction protocol which may help this problem: decrease the primer:template ratio to 1:1 or less, and carry out sequencing reactions at 16°C (Klenow only; Sequenase does not work well at lower temperatures). If the problem continues, try another primer sequence.

(ii)   Peaks in two or three or all four colours at one (or more) places in the data. This problem is usually sequence-related; there are two common causes First, there may be secondary structure in the template which the enzyme has difficulty reading through, leading to a false termination (not ending in a ddNTP). This problem is most common with Sequenase and least common with Taq polymerase.

If this has occurred in a Sequenase reaction, it can often be overcome by doubling the amount of enzyme used, or by doing a Klenow 'chase' with 1 unit of Klenow polymerase added (C reaction) for 3 min at the end of the Sequenase incubation. Klenow polymerase problems may be improved by doing the reactions at a higher temperature (42°C or 50°C in the worst cases). Some sequences are more easily read through by reverse transcriptase. Also, the addition of single-stranded binding protein to the sequencing reactions has been reported to help this problem (see Chapter 4, Section 2.2.8).

If the overlapping peaks occur in a GC-rich region, and are followed by a region of wide spacing between peaks, then 'band compression' in the gel is the likely problem. This is again due to secondary structure, at the 3' end of extension products, causing anomalous migration in the gel. This problem is often minimized by substituting 7-deaza-dGTP for regular dGTP in the dNTP mixes. Sequences not helped by 7-deaza-dGTP may become readable by using dITP, although this tends to introduce other problems. Running the gel at higher temperatures (i.e. 50−60°C) may also help, although resolution generally suffers. As a final resort, formamide gels have been reported to resolve the worst cases, but these gels are difficult and expensive to prepare and have poor resolution relative to the normal urea gels.

5.2 Weak signal

This common problem is obvious to identify in the raw data; instead of peaks and valleys there are virtually flat lines. Analysed data look noisy or small, and automatic calls cannot be made. The problem can be one of two types.

(i)   Overall weak signal. This is often due to inaccurate template quantitation, either through omitting the UV absorbance measurement or through a contaminating nucleic acid (e.g. chromosomal DNA, RNA) giving a falsely high reading. Check agarose gel for template purity and UV measurement and calculations for possible errors. Taq polymerase will generally give the strongest signal for a given amount of template. Another common cause is some problem in the sequencing reactions. Possibilities include the presence of enzyme inhibitors, a partially or completely inactive enzyme, concentration or purity problems with any sequencing reagent, or nuclease contamination. Check all reagents and enzymes to make sure they work on a known control template.

(ii) Declining signal throughout the run. This problem also has several different possible causes. A degraded template often gives this pattern, and generally gives noisy data or extra peaks even where the signal is visible. Check for a smeared band on an agarose gel.

Also suspect is the ethanol precipitation/resuspension/denaturation/loading process, with inadequate resuspension or denaturation of the larger fragments. Load a small quantity (2−3 μl) of the C reaction diluted in an equal volume of formamide/EDTA without precipitation to determine if the reaction proceeded as desired. Problems in this area are inadequate removal of ethanol following precipitation, overdrying of sample (may give overall low signal), incomplete resuspension, and incomplete denaturation of the sample before it enters the gel, due to inadequate heating. If there is a long time delay between heating and loading, another possibility is renaturation of the sample before electrophoresis into the gel.

Another possibility is an imbalance in the dNTP:ddNTP ratio. If the dNTP concentration is too low, the reaction will yield a higher concentration of short fragments relative to long ones. Conversely, if the ddNTP concentration is too low, long fragments will predominate, so the signal will initially be small and get progressively larger throughout the run. This is not likely to occur in all four of the reactions at once, but will generally be confined to a single reaction, and is therefore relatively easy to diagnose.

A final possibility is that the sample did not all follow the same pathway down the gel, but actually moved out of the detection area of the instrument. This can occur when the buffer leaks out of the upper buffer chamber, changing the electric field. Improved software that can 'follow' lanes of data more efficiently should mitigate problems of this type, and is currently under development.

5.3 **Poor resolution**

Poor resolution is usually associated with a gel problem. Uneven polymerization due to incomplete mixing of reagents, especially ammonium persulphate or TEMED, or poor quality reagents is one likely cause. Poor wells (tipped or wavy wells, or wells with residual acrylamide along the edges) may be indicated if the problem occurs in one or a few lanes, but not all. If the problem occurs regularly, check supplies for glass plates which are not flat, combs and/or spacers of varying thicknesses, or uneven contact between glass plates and the heat equalizing plate behind the gel. Clamping the upper buffer chamber too tightly can cause distortion of the wells and poor resolution.

The problem may be entirely unrelated to the gel. Impure primers and extended primers can run as broader bands, and degraded reactions often appear smeared. Again these things should be associated with a particular primer or reaction set(s).

6. SUMMARY

The area of automated DNA sequencing is being investigated from many different angles in both industry and academia throughout the world. Undoubtedly many improvements will be made in the next few years. Areas where advancements are likely include:

(i)   Increased throughput per instrument, accomplished through increased number of bases per lane, faster running conditions, and increased number of lanes per gel.

(ii)   Complete automation of the sequencing process, including automation of DNA cloning and purification procedures, and sequencing reactions.

(iii)   Pre-poured disposable gels, or re-usable gels.

(iv)   Improved software for DNA sequence determination, manipulation, and storage.

The advent of automated procedures has already greatly increased the number of bases that can potentially be sequenced per year. Anticipating further developments, researchers have begun to undertake huge sequencing projects such as the Human Genome Initiative. Molecular biology advancements in cloning and mapping will certainly accompany the developing automation. In the longer term, entirely new sequencing technologies may replace existing methods. The reader is advised to be aware of the latest developments, as procedures outlined here will soon be obsolete. Undoubtedly, an ability to determine even very large sequences quickly and reliably will have a profound impact on our understanding of biological processes. It is also quite possible that further progress in automating all of these procedures will mean that sequencing a gene may become as routine in disease diagnosis as growing up a bacterial culture is today.

7. REFERENCES

1. Maxam,A.M. and Gilbert,M. (1980) In *Methods in Enzymology.* Grossman,L. and Moldave,K. (eds), Academic Press, New York, Vol. 65, p. 499.
2. Sanger,F., Nicklen,S. and Coulson,A.R. (1977) *Proc. Natl. Acad. Sci. USA*, **74**, 5463.
3. Messing,J. (1983) In *Methods In Enzymology.* Wu,R., Grossman,L. and Moldave,K.(eds), Academic Press, New York, Vol. 101, p. 20.
4. Zagursky,R.J., Baumeister,K., Lomax,N. and Berman,M.L. (1985) *Gene Anal. Techn.*, **2**, 89.
5. Strauss,E.C., Kobori,J.A., Siu,G. and Hood,L.E. (1986) *Anal. Biochem.*, **154**, 353.
6. Henikoff,S. (1984) *Gene*, **28**, 351.
7. Tabor,S. and Richardson,C.C. (1987) *Proc. Natl. Acad. Sci. USA*, **84**, 4767.
8. Smith,L.M., Fung,S., Hunkapiller,M.W., Hunkapiller,T.J. and Hood,L.E. (1985) *Nucleic Acids Res.*, **13**, 2399.
9. Prober,J.M., Trainor,G.L., Dam,R.J., Hobbs,F.W., Robertson,C.W., Zagursky,R.J., Cocuzza,A.J., Jenson,A.M. and Baumeister,K. (1987) *Science*, **238**, 336.
10. Smith,L.M., Sanders,J.Z., Kaiser,R.J., Hughes,P., Dodd,C., Connell,C.R., Heiner,C., Kent,S.B.H. and Hood,L.E. (1986) *Nature*, **321**, 674.
11. Connell,C., Fung,S., Heiner,C., Bridgham,J., Chakarian,V., Heron,E., Jones,R., Menchen,S., Mordan,W., Raff,M., Recknor,M., Smith,L., Springer,J., Woo,S. and Hunkapiller,M. (1987) *Biotechniques*, **5**, 342.

INDEX